Lecture Notes in Mathematics

Edited by A. Dold and B. Eckmann

1281

Menegazzo G. Zacher (Eds.)

eory

nference held at
taly, May 25–31, 1986

Springer-Verlag
Berlin Heidelberg New York London Paris Tokyo

Editors

Otto H. Kegel
Mathematisches Institut, Albert-Ludwig-Universität
Albertstraße 23b, 7800 Freiburg, Federal Republic of Germany

Federico Menegazzo
Giovanni Zacher
Seminario Matematico dell' Università
Via Belzoni, 7, 35131 Padova, Italy

Mathematics Subject Classification (1980): 20-06

ISBN 3-540-18399-X Springer-Verlag Berlin Heidelberg New York
ISBN 0-387-18399-X Springer-Verlag New York Berlin Heidelberg

Printed in Germany

Printing and binding: Druckhaus Beltz, Hemsbach/Bergstr.
2146/3140-543210

Preface.

This volume presents the texts of most of the invited lectures and some of the contributions given at the International Conference on Group Theory which took place 26. - 31. May 1986 at Brixen/ Bressanone, Italy, on the premises of the Casa della Gioventù Universitaria dell'Università di Padova. There were 15 invited lecturers. In the afternoon 24 shorter communications were given. In all, there were 78 participants. Beside the formal lectures there were intensive informal contacts, discussions and exchanges of ideas, often till late in the night....

The conference was supported financially by the Ministero della Pubblica Istruzione, by the C.N.R., by the University of Padova and by the C.I.R.M. (Trento); the latter also assumed responsibility for the local organisation and administration.

We have to thank all the lecturers, participants, and the local organisers, whose enthusiasm and devotion made this meeting a success.

O. H. Kegel, Freiburg
F. Menegazzo, Padova
G. Zacher, Padova

List of Participants

1. Bernhard AMBERG (Mainz)
2. Marco BARLOTTI (Firenze)
3. Claudio BARTOLONE (Palermo)
4. Mariagrazia BIANCHI (Milano)
5. Arrigo BONISOLI (Modena)
6. Rolf BRANDL (Würzburg)
7. Brunella BRUNO (Padova)
8. Giorgio BUSETTO (Padova)
9. Andrea CARANTI (Trento)
10. Luisa CARINI (Messina)
11. Maria Rosaria CELENTANI (Napoli)
12. Gabriella CORSI TANI (Firenze)
13. Mario CURZIO (Napoli)
14. Alma D'ANIELLO (Mainz)
15. Rex DARK (Galway)
16. Francesco DE GIOVANNI (Napoli)
17. Lino DI MARTINO (Milano)
18. Walter DIRSCHERL (Würzburg)
19. Maurizio EMALDI (Padova)
20. Valeria FEDRI (Firenze)
21. Giovanni FERRERO (Parma)
22. Silvana FRANCIOSI (Napoli)
23. Alberto FRIGERIO (Padova)
24. Antonino GIAMBRUNO (Palermo)
25. Anna GILLIO BERTA MAURI (Milano)
26. Anna Luisa GILOTTI (Firenze)
27. Fletcher GROSS (Salt Lake City)
28. Karl GRUENBERG (London)
29. Narain GUPTA (Winnipeg)
30. Brian HARTLEY (Manchester)
31. Trevor HAWKES (Warwick)
32. Hermann HEINEKEN (Würzburg)
33. Wolfgang HERFORT (Wien)
34. Manfred J. KARBE (Berlin)
35. Otto KEGEL (Freiburg)
36. Enrico JABARA (Padova)
37. Hartmut LAUE (Kiel)
38. Felix LEINEN (Mainz)
39. John LENNOX (Cardiff)
40. Antonella LEONE (Napoli)
41. Patrizia LONGOBARDI (Napoli)
42. Andrea LUCCHINI (Padova)
43. Mercede MAJ (Napoli)
44. Ermanno MARCHIONNA (Milano)
45. Sandro MATTAREI (Trento)
46. Federico MENEGAZZO (Padova)
47. Martin MENTH (Würzburg)
48. Claudia METELLI (Padova)
49. Walter MÖHRES (Würzburg)
50. Carlo MORINI (Ferrara)
51. Franco NAPOLITANI (Padova)

52. Andreas NEUMANN (Trento)
53. Peter NEUMANN (Oxford)
54. Martin L. NEWELL (Galway)
55. Péter Pàl PALFY (Budapest)
56. Virgilio PANNONE (Firenze)
57. Giuseppe PIRILLO (Firenze)
58. Peter PLAUMANN (Erlangen)
59. Salvatore RAO (Napoli)
60. Derek J.S. ROBINSON (Urbana)
61. Antonio ROSATI (Firenze)
62. James ROSEBLADE (Cambridge)
63. Roland SCHMIDT (Kiel)
64. Benedetto SCIMEMI (Padova)
65. Carlo M. SCOPPOLA (Trento)
66. Steward STONEHEWER (Warwick)
67. Luigi SERENA (Firenze)
68. M. Clara TAMBURINI BELLANI (Milano)
69. Umberto TIBERIO (Firenze)
70. Cesarina TIBILETTI MARCHIONNA (Milano)
71. Sean TOBIN (Galway)
72. Libero VERARDI (Bologna)
73. Erich WALTER (Trento)
74. John WILSON (Cambridge)
75. Giovanni ZACHER (Padova)
76. Vittoria ZAMBELLI (Milano)
77. Guido ZAPPA (Firenze)
78. Irene ZIMMERMANN (Freiburg)

CONTENTS

FAITHFUL SIMPLE MODULES FOR THE NON-ABELIAN GROUP OF ORDER pq

Dedicated to Guido Zappa, on his 70th birthday, 7.12.1985

Marco Barlotti
Ist. Mat. "U. Dini"
Università di Firenze
viale Morgagni 67/a
I 50134 Firenze (Italy)

1. Introduction

We investigate (in section 2) a class of faithful simple modules for certain metacyclic groups; this leads to a complete description (in section 3) of the faithful simple modules for the non-abelian group of order pq (p,q primes) over a finite field of characteristic different from p and q. These results have been obtained aiming to a study of Fitting formations which will appear later but are, hopefully, of indipendent interest.

Throughout the sequel, t will be a fixed prime and s will be a fixed positive integer; F will denote the finite field with t^s elements. Further notation will be established within section 3.

2. Modules for a class of metacyclic groups

Definition 2.1 - Let m be a positive integer not divisible by t, and let a be the multiplicative order of t^s modulo m (i.e., a is the smallest positive integer such that m divides $t^{as}-1$). For any divisor d of a, we define

$$G_m^d = \langle x,y \;/\; x^m=y^d=1 \;,\; y^{-1}xy=x^r\rangle$$

where $r=t^{as/d}$.

Clearly, G_m^d is the semi-direct product of $\langle x\rangle$ (which is cyclic of order m) by $\langle y\rangle$ (which is cyclic of order d), hence has order md .

Theorem 2.2 - Let m be a positive integer not divisible by t , and let a be the multiplicative order of t^s modulo m ; let d be a divisor of a .

Let K be the field with t^{as} elements, let u be an element of multiplicative order m in K , and let V be the additive group of K with the natural structure of F-vector space; let G be the group G_m^d defined in 2.1 .

(a) There is an action of G on V such that, for all $v\in V$, $vx=vu$ (i.e., vx is the product in K of v by u) and $vy=v^{t^{as/d}}$ (i.e., vy is v raised to the $t^{as/d}$-th power in K).

(b) Under the action considered in (a) , V is a faithful simple FG-module, which we shall denote by V_u .

(c) $Hom_{FG}(V_u,V_u)$ is the field with $t^{as/d}$ elements; in particular, if $d=a$ then V_u is a faithful absolutely irreducible FG-module.

Proof - It is easy to verify (a), and it is well known (see, e.g., [2] II 3.10) that V is then faithful and simple for $\langle x\rangle$, which implies (b) since $\langle x\rangle$ contains every minimal normal subgroup of G .

To prove (c), we first consider the case $d=1$ (note that now G is cyclic of order m). Let $H = Hom_{FG}(V_u,V_u)$. Multiplication by a given element of K is clearly a FG-endomorphism of V , so H contains a subfield isomorphic to K ; because H is a field ([2],I 10.5), $|H|=t^{ash}$ for a convenient positive integer h : we must show that $h=1$. Let H^o be the multiplicative group of H : clearly, V_u is a faithful FH^o-module; but V also is simple for H^o , because multiplication by u generates a subgroup of H^o

which acts on V_u exactly as G does. So, by [2] II 3.10 , $t^{as} \equiv 1$ modulo $t^{ash}-1$, which implies that $h=1$.

Now we can consider the general case, where d is any divisor of a . Let f be a FG-endomorphism of V_u : since f is in particular a $F\langle x\rangle$-endomorphism of V_u , we have just shown that there exists an element w of K such that f acts via multiplication by w in K ; moreover, $vyf=vfy$ for every $v\in V$. This implies that $(vy)w=(vw)y$, i.e.,

$$v^{t^{as/d}}w = (vw)^{t^{as/d}} = v^{t^{as/d}}w^{t^{as/d}}$$

whence $w^{t^{as/d}} = w$, since v can be taken to be different from zero. Thus w belongs to the subfield K_o of K of order $t^{as/d}$, so that $Hom_{FG}(V_u,V_u)$ has at most $t^{as/d}$ elements; on the other hand, it is easy to check that multiplication by a given element of K_o is a FG-endomorphism of V_u , so we are done.

3. Modules for the non-abelian group of order pq

We fix some further notation throughout this section. Let p,q be primes different from t such that q divides $p-1$; let a be the multiplicative order of t^s modulo p , and denote by K the field with t^{as} elements (an extension of F). Let G be the non-abelian group of order pq , let $P=\langle x\rangle$ be its p-Sylow subgroup and let y be an element of order q in G .

Lemma 3.1 - If there exists a faithful simple FG-module V which is simple for P , then q divides a .

Proof - Since V is a faithful simple FP-module, we may suppose w.l.o.g. that V is the additive group of K with the natural structure of F-vector space and that there exists an element u of multiplicative order p in K such that vx is the product vu in K for every $v\in V$ (see, e.g., [2] II 3.10).

Let m be a generator of the multiplicative group of K, and denote by f the F-automorphism of V such that $vf=vy$ for every $v\in V$; there exists a positive integer $h<t^{as}$ such that

$$(mf)(1f)^{-1} = m^h \tag{1}$$

where 1 denotes the multiplicative unit in K.

Denote by g the mapping $V\to V$ such that $vg=v^h$ for every $v\in V$. We shall show that

(a) g is a field automorphism of K over F, and

(b) g has order q.

The assertion of the lemma will then follow from [3] VII §5.

To prove (a) and (b), we need a preliminary observation about f and g, i.e.,

$$vg = (vf)(1f)^{-1} \quad \text{for every } v\in V . \tag{2}$$

[Proof of (2) - Since V is a simple FP-module, for every $w\in V$ there exist $c_0, c_1, .., c_{p-1}\in F$ such that

$$w = c_0+c_1u+c_2u^2+\ldots+c_{p-1}u^{p-1}$$

whence

$$wf = c_0(1f)+c_1(uf)+\ldots+c_{p-1}(u^{p-1}f) \quad .$$

There exists a primitive q-th root r of 1 modulo p such that $x^iy=yx^{ir}$ in G for every positive integer i; this implies that for every $v\in V$ and for every positive integer i

$$(vu^i)f = (vx^i)y = (vy)x^{ir} = (vf)u^{ir} \quad . \tag{3}$$

Thus
$$\begin{aligned} wf &= c_0(1f)+c_1u^r(1f)+\ldots+c_{p-1}u^{r(p-1)}(1f) = \\ &= (1f)(c_0+c_1u^r+\ldots+c_{p-1}u^{r(p-1)}) . \end{aligned}$$

Hence for all $v,w\in V$ we have, using (3) again,

$$\begin{aligned} (vw)f &= (v(c_0+c_1u+\ldots+c_{p-1}u^{p-1}))f = \\ &= (c_0v+c_1vu+\ldots+c_{p-1}vu^{p-1})f = \\ &= c_0vf+c_1vuf+\ldots+c_{p-1}vu^{p-1}f = \end{aligned}$$

$$= c_0 vf + c_1 vfu^r + \ldots + c_{p-1} vfu^{r(p-1)} =$$
$$= (vf)(c_0 + c_1 u^r + \ldots + c_{p-1} u^{r(p-1)}) =$$
$$= (vf)(wf)(1f)^{-1}$$

i.e.,

$$(vw)f = (vf)(wf)(1f)^{-1} \quad . \tag{4}$$

Using (1) and (4), it is easy to prove by induction on i that $m^{ih} = (m^i f)(1f)^{-1}$ for every positive integer i, which implies (2) since every non-zero element of K is a power of m.]

Now we proceed to prove (a). By (2), g is one-to-one since such is f. Still by (2), for all $v, w \in V$

$$(v+w)g = ((v+w)f)(1f)^{-1} = (vf+wf)(1f)^{-1} =$$
$$= (vf)(1f)^{-1} + (wf)(1f)^{-1} = vg + wg$$

and it is clear by definition that $(vw)g = (vg)(wg)$ for all $v, w \in V$. There remains to show that g leaves every element of F fixed; in fact, if $e \in F$, $eg = (ef)(1f)^{-1}$ by (2); but $ef = (e1)f = e(1f)$ since f is a F-automorphism of V, hence $eg = e$ and we are done.

At last we prove (b). Using (2) and (4), it is easily seen by induction on i that $vg^i = (vf^i)(1f^i)^{-1}$ for all $v \in V$ and for every positive integer i, whence $vg^q = v$ for all $v \in V$ since f has clearly order q. Thus the order of g divides q, and all we have to do is show that g is not the identity mapping. In fact, $ug = u$ would imply by (2) that $uf = (1f)u$, while by (3) we know that $uf = (1f)u^r$ and $u^r \neq u$ because u has multiplicative order p while r is a primitive q-th root of 1 modulo p.

Lemma 3.2 - Let G_0 be a group with a self-centralizing normal subgroup A such that every subgroup of G_0 containing A is normal in G, and suppose that t does not divide $|G_0|$.

Let W be the right regular FG_0-module, and let (by Maschke's theorem) $W = W_1 + \ldots + W_h$ where each W_i is a simple FG_0-module.

Then any faithful simple FG_0-module appears exactly $|G:A|$ times among the W_i's.

Proof - This result is contained in the proof of Lemma 4.2 in [1] (that lemma was stated for $s=1$, but such restriction is clearly unnecessary).

Theorem 3.3 - If q divides a (whence $G=G_p^q$ in the notation of 2.1) then any faithful simple module for G is one of the faithful simple modules for G_p^q described in 2.2 .

Proof - Let $n=(p-1)/a$. It is well known that there exist exactly n pairwise non-isomorphic faithful simple FP-modules, each of dimension a; they give rise (as described in 2.2) to n pairwise non-isomorphic faithful simple FG-modules $V_1, .., V_n$, each of dimension a over F.

Now consider the right regular FG-module W. It must be

$$W = W_0 \oplus W_1 \oplus \dots \oplus W_n \oplus \overline{W}$$

where:

- W_0 is a direct sum of representatives of the isomorphism classes of simple FG-modules with P in the kernel (hence has dimension q over F);

- each W_i (for $i=1,..,n$) is (by 3.2) a direct sum of q FG-isomorphic copies of a given faithful simple FG-module V_i among those considered above (hence has dimension qa over F);

- any possible further simple FG-module is FG-isomorphic to a FG-submodule of $\overline{W}$.

Since W has dimension $pq=q+nqa$ over F, we deduce that $\overline{W}=0$, i.e. that no other simple FG-module exists.

Theorem 3.4 - If q does not divide a, then

(a) any FG-module induced from a faithful simple FG-module is faithful and simple for G;

(b) any faithful simple FG-module is induced from a faithful simple FP-module.

Proof of (a) - Let V be a faithful simple FP-module. The induced module V^G is a direct sum

$$V^G = (V\otimes 1) \oplus (V\otimes y) \oplus \dots \oplus (V\otimes y^{q-1}) \tag{5}$$

where each $V\otimes y^i$ is a faithful simple FP-module (not necessarily FP-isomorphic to V).

Now let W be a simple FG-submodule of V^G . If W were not faithful for G , then it would be a trivial FP-module: but, by (5) and the theorem of Jordan-Holder, V^G does not possess trivial FP-submodules; hence, W is a faithful simple FG-module.

If W were a homogeneous FP-module, then by [1] Lemma 4.2 it would be simple: but this is impossible by 3.1 because q does not divide a . So, by Clifford's theorem ([2] V 17.3), if W_0 is a simple FP-submodule of W and V_0 is the sum of all the FP-submodules of W which are FP-isomorphic to W_0 , then $W = V_0^G$.

Since, as we have already remarked, V^G does not possess trivial FP-submodules, W_0 must be faithful and therefore has dimension a over F ([2] II 3.10). So

$$\dim_F W = q \dim_F V_0 \geqslant q \dim_F W_0 = qa = \dim_F V^G$$

whence $W=V^G$ and (a) is proved.

Proof of (b) - By (a) , every faithful simple FP-module induces a faithful simple FG-module: but, of course, non-isomorphic FP-modules may induce isomorphic FG-modules.

Since q does not divide a but divides $p-1$, the number $(p-1)/a$ of pairwise non isomorphic faithful simple FP-modules is a multiple of q : let $n=(p-1)/aq$. The $(p-1)/a$ pairwise non-isomorphic faithful simple FP-modules, each of dimension a over F , induce (at least) n pairwise non-isomorphic faithful simple FG-modules $V_1, .., V_n$, each of dimension qa over F .

Now consider, as in the proof of 3.3 , the right regular FG-module W . It must be

$$W = W_0 \oplus W_1 \oplus \dots \oplus W_n \oplus \overline{W}$$

where:

- W_o is a direct sum of representatives of the isomorphism classes of simple FG-modules with P in the kernel (hence has dimension q over F);

- each W_i (for $i=1,..,n$) is (by 3.2) a direct sum of q FG-isomorphic copies of a given faithful simple FG-module V_i among those considered above (hence has dimension q^2a over F);

- any possible further simple FG-module is FG-isomorphic to a FG-submodule of $\overline{W}$.

Since W has dimension $pq=q+nq^2a$ over F , we deduce that $\overline{W}=0$, i.e. that no other simple FG-module exists.

Corollary 3.5 - All the faithful simple FG-modules have the same dimension over F (i.e., the least common multiple of q and a).

Proof - Clear from 3.3 and 3.4 .

REFERENCES

[1] T. O. Hawkes, ON THE AUTOMORPHISM GROUP OF A 2-GROUP, Proc. Lond. Math. Soc. (3) 26 (1973) 207-225

[2] B. Huppert, ENDLICHE GRUPPEN I, Springer-Verlag 1967

[3] S. Lang, ALGEBRA, Addison-Wesley 1965

Some examples of infinite groups in which each element commutes with its endomorphic images†

A. Caranti††
Mathematisches Institut
Universität Erlangen-Nürnberg
Bismarckstraße $1^1/_2$
D-8520 Erlangen
Federal Republic of Germany

S. Franciosi, F. de Giovanni
Dipartimento di Matematica "R. Caccioppoli"
Università di Napoli
via Mezzocannone 8
I-80134 Napoli
Italy

1. Introduction

An E-group is a group G in which each element commutes with all of its images under endomorphisms,

$$[a,a\varphi] = 1, \text{ for all } a \in G, \varphi \in \mathrm{End}G.$$

The interest for E-groups arose because they are exactly the groups in which the endomorphisms do generate a ring, i.e. the groups in which the set of maps on G

$$\{\varphi_1 + \ldots + \varphi_n \mid \varphi_i \in \mathrm{End}G\}$$

forms a ring under the natural operations: see [3] and [1] for this and other general facts on the subject. We only recall here that an E-group G is obviously a 2-Engel group, and hence it is nilpotent of class at most 3 and has $\gamma_3(G)^3 = 1$ (see [5] Part 2, p. 45-46). No example of class 3 being actually known, we will be concerned in the sequel with E-groups of nilpotency class 2. However, note that E-groups without elements of order 3 have nilpotency class at most 2, and that the same holds for E-groups of exponent p^2 (see [3]).

In this paper we construct the first examples of infinite E-groups, showing that the situation in this case does not differ from the finite one, at least as far as automorphisms are concerned, as explained in the following discussion.

---

† Partially supported by MPI, Italy

†† Current address: Dipartimento di Matematica, Università di Trento, I-38050 Povo (Trento), Italy

The simplest way of getting an E-group is to construct a group G with the properties:
(1.1) every endomorphism of G which is not an automorphism maps G into Z(G), and
(1.2) every automorphism of G is central (i.e. acts as the identity on G/Z(G)).

It is shown in [1] that (1.1) holds in a finite E-group G if and only if G is directly indecomposable. However, the main tool for proving the "if" part is Fitting's lemma, which fails in general for infinite groups. We are not able to decide what happens in the general case. All our infinite examples here do in fact satisfy (1.1).

As to (1.2), in [1] it is shown that for a finite E-group AutG need not coincide with its normal subgroup $\mathrm{Aut}_c G$ consisting of the central automorphisms, but the factor group $\mathrm{AutG}/\mathrm{Aut}_c G$ has to be abelian. The same holds also for infinite E-groups. Here we provide among others examples supporting the first statement; as to the second one, the proof of [1] applies in the general case.

The paper is organized as follows.

In §2 torsion-free examples are constructed. The first one is already contained in [2], and has $\mathrm{AutG} = \mathrm{Aut}_c G$. The same homological techniques of Fournelle allow us to construct another examples with $\mathrm{AutG}/\mathrm{Aut}_c G$ cyclic of order 3. We are grateful to H. Heineken for remarking that it would have been indeed possible to mix the ideas of Fournelle with construction methods typical of the finite case.

In §3 torsion examples, both bounded and unbounded, are constructed. These examples require much more work than the torsion-free ones, in which the homological techniques play a simplifying role.

Our notation is standard, as in [6].

Acknowledgments

This paper was first conceived in Erlangen, during a visit of the last two authors, and actually written down in Napoli, during a visit of the first author. We would like to thank all the people that made this possible, in particular Karl Strambach and Mario Curzio for their warm hospitality.

2. Torsion-free examples

In this section we construct two torsion-free examples of non-abelian E-groups. It can easily be shown, analogously to Theorem 1 of [3], that an E-group that admits a factor group isomorphic to $\mathbb{Z}$ has to be abelian. Therefore it is not surprising that the building blocks for the examples that follow are taken to be rational groups.

2.1. Our first example is already contained in [2], but we recall it here briefly for the convenience of the reader.

Take Λ_1, Λ_2 to be two infinite disjoint sets of primes. Consider the rational groups

$$X_1 = \langle {}^1/_p \mid p \in \Lambda_1 \rangle,$$
$$X_2 = \langle {}^1/_p \mid p \in \Lambda_2 \rangle,$$
$$C = \langle {}^1/_p \mid p \in \Lambda_1 \cup \Lambda_2 \rangle$$

and let x_1, x_2, c represent 1 in X_1, X_2, C respectively. Write $Q = X_1 \times X_2$.

It is proved in [2] that if $A \leq B$ are rational groups containing 1, with the property that there are infinitely many primes p such that ${}^1/_p \in A$ but the p-height of 1 in B is finite, then Ext(A,B) is non-trivial, torsion-free and divisible. This applies here for B = C and $A = X_i$, i = 1, 2.

Let $\Delta \in H^2(Q,C)$, and choose a 2-cocycle $f \in \Delta$. Define a 2-cocycle f' by f'(x,y) = f(y,x), and a class $\Delta' \in H^2(Q,C)$ by $\Delta' = [f']$.

As in [2], we define a homomorphism

$$H^2(Q,C) \rightarrow \mathrm{Ext}(Q,C)$$
$$\Delta \rightarrow \Delta + \Delta'.$$

This homomorphism leads canonically to a splitting of the sequence of the Universal Coefficient Theorem, because it restricts on Ext(Q,C) to the multiplication by 2, which is an automorphism of Ext(Q,C), since this latter group is 2-divisible and has no elements of order 2. This splitting is natural, as shown in Proposition 2.1 of [2].

To define a central extension

$$C \rightarrowtail G \twoheadrightarrow Q$$

we therefore assign

(i) the commutator form, i.e. an alternating bilinear map $Q \times Q \rightarrow C$, defined by

$$c = [x_1, x_2];$$

(ii) an element

$$(\sigma_1, \sigma_2) \in \mathrm{Ext}(Q,C) = \mathrm{Ext}(X_1,C) \oplus \mathrm{Ext}(X_2,C),$$

where $\sigma_1 \neq 0$, $\sigma_2 \neq 0$.

It is clear that endomorphisms of C are multiplications by elements $t \in \mathbb{Z}$, and those of Q are multiplications by $r \in \mathbb{Z}$ on X_1, $s \in \mathbb{Z}$ on X_2, since X_1, X_2 are of incomparable types. According to [2], such a pair of endomorphisms induces an endomorphism of G if and only if

$$tc = [ra_1, sa_2]$$

and

$$(r\sigma_1, s\sigma_2) = t(\sigma_1, \sigma_2),$$

i.e. t = rs, and r = rs = s. Thus, either r = s = 0, so that $G\varphi \leq Z(G)$, or r = s = 1, and $\varphi \in \mathrm{Aut}_c G$.

Thus this example satisfies (1.1) and (1.2).

2.2. Our second torsion-free example is an E-group G with $\mathrm{Aut}G \neq \mathrm{Aut}_c G$.

Let Λ_1, Λ_2, Λ_3 be infinite pairwise disjoint sets of prime numbers, and for i = 1, 2, 3 denote by X_i, Y_i two copies of the rational group

$$\langle {}^1/_p \mid p \in \Lambda_i \rangle$$

and by Z_i, W_i two copies of the subgroup

$$\langle {}^1/_p \mid p \in \Lambda_j \cup \Lambda_k \rangle$$

where $\{i,j,k\} = \{1,2,3\}$. Denote the element 1 of X_i, Y_i, Z_i, W_i by x_i, y_i, z_i, w_i respectively, and write $C = \mathrm{Dr}_{i=1}^{3}(Z_i \times W_i)$, $Q = \mathrm{Dr}_{i=1}^{3}(X_i \times Y_i)$.

We define a central extension

$$C \rightarrowtail G \twoheadrightarrow Q$$

by assigning, as in 2.1,

(i) the commutator map defined by

$$[x_2,x_3] = z_1\,,\ [x_3,x_1] = z_2\,,\ [x_1,y_2] = z_3\,,$$

$$[y_2,y_3] = w_1\,,\ [y_3,y_1] = w_2\,,\ [x_1,x_2] = w_3\,,$$

$$[x_i,y_j] = [y_i,x_j] = -[x_i,x_j] - [y_i,y_j] \text{ if } i \neq j,\ [x_i,y_i] = 1 \text{ for all } i, \text{ and}$$

(ii) an element $\Sigma \in \mathrm{Ext}(Q,C)$ whose only non-zero components are $\sigma_{ij} \in \mathrm{Ext}(X_i,Z_j)$, $\tau_{ij} \in \mathrm{Ext}(Y_i,W_j)$, for $i \neq j$, with $\sigma_{ij} = \tau_{ij}$.

Since C/G' is the torsion subgroup of G_{ab}, C is a fully invariant subgroup of G. Therefore every endomorphism τ of G acts on C and Q, and the incomparability of the chosen types forces τ to map the subgroups $X_i \times Y_i$ and $Z_i \times W_i$ ($i = 1, 2, 3$) into themselves. Let A be the rational matrix which represents the actions of τ on $Z_3 \times W_3$ with respect to the elements z_3, w_3, and let A_i, $i = 1, 2, 3$, be the rational matrices which represent the action τ on $X_i \times Y_i$ with respect to the elements x_i, y_i. We have, since $\tau \in \mathrm{End}G$, $A_i D_{i3} = D_{i3} A_i$, where

$$D_{i3} = \begin{bmatrix} \sigma_{i3} & 0 \\ 0 & \tau_{i3} \end{bmatrix}$$

is the component of Σ in $\mathrm{Ext}(X_i \times Y_i, Z_3 \times W_3)$, $i = 1, 2$. Since $\sigma_{i3} = \tau_{i3}$ has infinite order, $A_i = A$ and so $A_1 = A = A_2$. The same argument shows that τ acts in the same way on all the subgroup $X_i \times Y_i$ and $Z_i \times W_i$.

Let

$$M = \begin{bmatrix} r & s \\ t & u \end{bmatrix}$$

be the rational matrix that represents this common action. From the commutator relations in G we obtain the equalities

$$\begin{cases} r = r^2 - 2rs \\ s = s^2 - 2rs \\ t = t^2 - 2tu \\ u = u^2 - 2tu \end{cases} \qquad \begin{cases} r = ru + st - su \\ s = rt - su \\ t = 2rs - s^2 \\ u = r^2 - s^2 \end{cases}$$

we obtain easily that either $M = 0$, or $M \in \langle N \rangle$, where

$$N = \begin{bmatrix} 0 & 1 \\ -1 & -1 \end{bmatrix}$$

It follows that G satisfies (1.1), and that AutG = $Aut_cG <\gamma>$, where γ is an automorphism of G which induces N on Q, and therefore $|AutG/Aut_cG| = 3$.

It can easily be verified, using the defining relations (i), that $[g,g\gamma] = 1$ for all $g \in G$. This completes the proof that G is an E-group with the announced properties.

3. Torsion examples

In this section we present several examples of non-abelian torsion E-groups. The following observation shows that our examples are quite general.

Remark. Let G be a non-abelian p-group which is an E-group, and let D be the maximum radicable subgroup of G. Then G/D has finite exponent. In particular, G has finite exponent if it is reduced.

Proof. Since G cannot have cyclic quotients of arbitrarily large order ([3] Theorem 1), the group G_{ab} is the direct product of a group of finite exponent and of a radicable group. The nilpotency of G implies that G' has finite exponent [4]; the group G/D is reduced and so G/G'D is also reduced and hence it has finite exponent, so that G/D has finite exponent.

3.1. The first example of this section is a countably infinite non-abelian E-group G of exponent p^2 (p odd) with $AutG = Aut_cG$.

Let G be the nilpotent p-group of class 2 generated by the sequence $(x_i)_{i \in \mathbb{Z}}$ of elements subject to the additional relations

$$x_{2i}{}^p = [x_{2i}, x_{2i+1}],\ x_{2i+1}{}^p = [x_{2i-1}, x_{2i+2}],\ [x_1,x_5] = 1.$$

Then G has exponent p^2 and $G' = Z(G) = \Omega_1(G)$. Write $C = Z(G)$, $Q = G/C$ and $v_i = x_iC$, for all i. Then Q and C are GF(p)-vector spaces, which we write additively, and have basis

$$\{v_i \mid i \in \mathbb{Z}\} \text{ and } \{[v_i, v_j] \mid i,j \in \mathbb{Z},\ i < j,\ (i,j) \neq (1,5)\},$$

where we use the fact that commutators depend only on the coset modulo Z(G). Now the commutator map is a bilinear form $Q \times Q \to C$, and the p-th power map is a homomorphism $Q \to C$.

First we will find the solutions of the equation

$$u^p = [u,v] \tag{3.1}$$

where u and v are non-trivial elements of Q, and v^p is a commutator in G. Let $u = \Sigma_i\, \alpha_i v_i$ and $v = \Sigma_i\, \beta_i v_i$, where $\alpha_i, \beta_i \in GF(p)$. Since $u^p = \Sigma_i\, \alpha_i\, v_i^p$, equation (3.1) becomes

$$\Sigma_i\, \alpha_{2i}\, [v_{2i}, v_{2i+1}] + \Sigma_i\, \alpha_{2i+1}\, [v_{2i-1}, v_{2i+2}] = \\ = \Sigma_{i<j}\, (\alpha_i\beta_j - \alpha_j\beta_i)[v_i, v_j] \tag{3.2}$$

If $\alpha_{2i+1} \neq 0$ for some i, then $\alpha_{2i-1}\beta_{2i+2} - \alpha_{2i+2}\beta_{2i-1} \neq 0$, so that the (2i-1)-th and (2i+2)-th columns of the infinite matrix

$$M = \begin{bmatrix} \alpha_i \\ \beta_i \end{bmatrix}, i \in \mathbb{Z}.$$

are linearly independent; then $\alpha_j = \beta_j = 0$ for all $j \notin \{2i-2, 2i-1, 2i+2, 2i+3, 1, 5\}$, and hence $2i+1 = 1$ or $2i+1 = 5$, and this is impossible.

Thus $\alpha_{2i+1} = 0$ for all i and equation (3.2) becomes

$$\Sigma_i \alpha_{2i} [v_{2i}, v_{2i+1}] = \Sigma_{i<j} (\alpha_i\beta_j - \alpha_j\beta_i)[v_i, v_j] \qquad (3.3)$$

Therefore there exists an i such that $\alpha_{2i} \neq 0$, and so the 2i-th and (2i+1)-th columns of M are linearly independent. From this it follows that $\alpha_{2j} = \beta_{2j} = 0$ for each $j \neq i$, and since

$$\det \begin{bmatrix} \alpha_{2i+1} & \alpha_{2j+1} \\ \beta_{2i+1} & \beta_{2j+1} \end{bmatrix} = 0,$$

we obtain also $\beta_{2j+1} = 0$. Then $u = \alpha_{2i} v_{2i}$ and $v = \beta_{2i} v_{2i} + \beta_{2i+1} v_{2i+1}$, so that $\beta_{2i+1} = 1$. Write $v^p = [a,b]$, for suitable $a = \Sigma_i \gamma_i v_i$, $b = \Sigma_i \delta_i v_i$, $(\gamma_i, \delta_i \in GF(p))$, we have

$$\beta_{2i} [v_{2i}, v_{2i+1}] + [v_{2i-1}, v_{2i+2}] = = \Sigma_{i<j} (\gamma_i\delta_j - \gamma_j\delta_i)[v_i, v_j] \qquad (3.4)$$

Then $\gamma_{2i-1}\delta_{2i+2} - \gamma_{2i+2}\delta_{2i-1} = 1$ and $\gamma_{2i}\delta_{2i+2} - \gamma_{2i+2}\delta_{2i} = \gamma_{2i-1}\delta_{2i} - \gamma_{2i}\delta_{2i-1} = 0$, so that $\gamma_{2i} = \delta_{2i} = 0$, and

$$\beta_{2i} = \det \begin{bmatrix} \gamma_{2i} & \gamma_{2i+1} \\ \delta_{2i} & \delta_{2i+1} \end{bmatrix} = 0,$$

Therefore we have $u = \alpha_{2i} v_{2i}$ and $v = v_{2i+1}$ for some i.

Let τ be an endomorphism of G, and assume that $G\tau \nsubseteq C$; then $x_j\tau \notin C$ for some j, and it follows easily that $x_j\tau \notin C$ for all j.

From the relations we obtain $(v_{2j}\tau)^p = [v_{2j}\tau, v_{2j+1}\tau]$ and hence $v_{2j}\tau = \alpha v_{2k}$ $(\alpha \in GF(p))$ and $v_{2j+1}\tau = v_{2k+1}$ for some k. Then for each j we can find integers h, k, e such that $[v_{2h-1}, \alpha v_{2k+2}] = [v_{2j-1}\tau, v_{2j+2}\tau] = (v_{2j+1}\tau)^p = v_{2e+1}{}^p = [v_{2e-1}, v_{2e+2}]$, and therefore $h = k = e$, and $\alpha = 1$; thus the number $d = 2(k-j)$ is the same for all j, and $v_i\tau = v_{i+d}$ for all i. Since $[v_{1+d}, v_{5+d}] = 1$, we have $d = 0$, and τ acts as the identity map on Q. It follows that G is an E-group.

3.2. This second example G has again exponent p^2 but $AutG/Aut_cG$ of oder 3.

Let p be a prime such that $p \equiv 1 \pmod 3$ and let G be the nilpotent p-group of class 2 generated by two sequences $(x_i)_{i \in \mathbb{Z}}$, $(y_i)_{i \in \mathbb{Z}}$ of elements with the additional relations

$$x_i{}^p = [y_i, y_{i+1}],\ y_{2i}{}^p = [x_{2i}, x_{2i+1}],\ y_{2i+1}{}^p = [x_{2i-1}, x_{2i+2}],$$
$$[x_i, x_{i+5}] = [x_i, x_{i+6}],\ [x_i, y_j] = 1,\ [x_0, x_4] = 1\ (i,j \in \mathbb{Z}).$$

Then G has exponent p^2 and $G' = Z(G) = \Omega_1(G)$.

If ω is a primitive third root of 1 in GF(p), the positions

$$x_i\varphi = x_i^{\omega}\ , \ y_i\varphi = y_i^{\omega^2}\ ,$$

for $i \in \mathbb{Z}$ define a non-central automorphism of G such that $[a,a\varphi] = 1$ for each $a \in G$. Write $C = Z(G)$, $Q = G/C$ and $u_i = x_iC$, $v_i = y_iC$ for all i. Then C and Q are GF(p)-vector spaces with basis

$$\{[u_i , u_j], [v_h , v_k] \mid i < j, j \neq i+6, (i,j) \neq (0,4), h < k\}$$

and

$$\{u_i , v_i \mid i \in \mathbb{Z}\}$$

respectively, and we will use additive notation for them. Now the commutator map is a bilinear form $Q \times Q \to C$, and the p-th power map is a homomorphism $Q \to C$.

Let σ be an endomorphism of G and denote by τ the induced homomorphism $G \to Q$. We may assume $\tau \neq 0$, so that either $x_j\tau \neq 0$ or $y_j\tau \neq 0$ for some j, and it follows easily that $x_i\tau \neq 0$ and $y_i\tau \neq 0$ for all i. We will prove in eleven steps that each element of G commutes with its image under σ.

Consider the following condition:

$$x_i\tau \in \langle v_j \mid j \in \mathbb{Z}\rangle \text{ and } y_h\tau \in \langle u_k \mid k \in \mathbb{Z}\rangle \text{ for all } i,h \in \mathbb{Z} \qquad (3.5)$$

(Step 1) If (3.5) holds, then $x_i\tau = \alpha_r v_r + \alpha_{r+1}v_{r+1} + \alpha_{r+2}v_{r+2}$ and $y_h\tau = \gamma_s u_s + \gamma_{s+1}u_{s+1}$, where the coefficients belong to GF(p).

For each i there exist h, t such that $y_h^p = [x_i , x_t]$; write $x_i\tau = \Sigma_j\ \alpha_j v_j$, $x_t\tau = \Sigma_q\ \beta_q v_q$, $y_h\tau = \Sigma_k\ \gamma_k u_k$. Then $\Sigma_k\ \gamma_k\ [v_k , v_{k+1}] = \Sigma_k\ \gamma_k u_k^p = (y_h\tau)^p = [x_i\tau , x_t\tau] = \Sigma_{j<q}\ (\alpha_j\beta_q - \alpha_q\beta_j)[v_j ,v_q]$ and hence $\alpha_j\beta_q - \alpha_q\beta_j = 0$ if $|j - q| \neq 1$. There exists r such that $\alpha_r\beta_{r+1} - \alpha_{r+1}\beta_r \neq 0$, and the claim follows easily.

(Step 2) If (3.5) holds, then $x_i\tau = \alpha_r v_r + \alpha_{r+1}v_{r+1}$ and $y_h\tau = \gamma_s u_s$, where the coefficients belong to GF(p).

Since $(x_i\tau)^p = [y_i\tau , y_{i+1}\tau]$, it follows that at least one of the coefficients α_r , α_{r+1} , α_{r+2} is zero and we may assume $x_i\tau = \alpha_r v_r + \alpha_{r+1}v_{r+1}$. From this it follows easily that $y_h\tau = \gamma_s u_s$.

(Step 3) If (3.5) holds, then $x_i\tau = \alpha_r v_r$ for some r.

Write $x_i\tau = \alpha_r v_r + \alpha_{r+1}v_{r+1}$, $y_i\tau = \gamma_s u_s$, $y_{i+1}\tau = \delta_t u_t$.

Then $\alpha_r v_r^p + \alpha_{r+1}v_{r+1}^p = (x_i\tau)^p = \gamma_s\delta_t\ [u_s , u_t]$ and the claim follows.

(Step 4) If (3.5) holds and $y_{2i}\tau \in \langle u_k\rangle$, then $x_{2i}\tau \in \langle v_k\rangle$, $x_{2i+1}\tau \in \langle v_k\rangle \cup \langle v_{k+1}\rangle$, $\{x_{2i-1}\tau, x_{2i+2}\tau\} \subseteq \langle v_{k+1}\rangle \cup \langle v_{k+2}\rangle$ and $k = 2j$ for some j.

If $y_{2i}\tau = \gamma_k u_k$, we have $\gamma_k\ [v_k , v_{k+1}] = (y_{2i}\tau)^p = [x_{2i}\tau , x_{2i+1}\tau]$, so that $\{x_{2i}\tau , x_{2i+1}\tau\} \subseteq \langle v_k\rangle \cup \langle v_{k+1}\rangle$. Since $\gamma_k[u_k , y_{2i+1}\tau] = (x_{2i}\tau)^p$, it follows that k is even and $x_{2i}\tau \in \langle v_k\rangle$, so that $y_{2i+1}\tau \in \langle u_{k+1}\rangle$.

We have also that $\{x_{2i-1}\tau, x_{2i+2}\tau\} \subseteq \langle v_{k+1}\rangle \cup \langle v_{k+2}\rangle$ since $[x_{2i-1}\tau, x_{2i+2}\tau] \in \langle x_{k+1}^p\rangle$.

(Step 5) Condition (3.5) does not hold.

If (3.5) holds, then $y_{2i}\tau \in \langle u_k\rangle$ and $y_{2i+2}\tau \in \langle u_q\rangle$ for some k, q, and actually $q = k + 2$. By Step 4, it follows that $x_{2i+1}\tau \in \langle v_{k+3}\rangle \cup \langle v_{k+4}\rangle$, which is impossible.

(Step 6) $x_i\tau \in \langle u_j \mid j \in \mathbb{Z}\rangle$ and $y_h\tau \in \langle v_j \mid j \in \mathbb{Z}\rangle$.

Suppose that $x_i\tau = \Sigma_j\ \alpha_j u_j + \beta_1 v_{t_1} + \dots + \beta_r v_{t_r}$, where $\beta_q \neq 0$ for $1 \leq q \leq r$. Since $[x_i\tau , y_h\tau] = 1$, we obtain $y_h\tau = \Sigma_j\ \gamma_j u_j + \delta_1 v_{t_1} + \dots + \delta_r v_{t_r}$, where either $\delta_q = 0$ for all $1 \leq q \leq r$, or $\beta_m\beta_n^{-1} = \delta_m\delta_n^{-1}$ for all m, n. It follows that $(x_k\tau)^p = [y_k\tau , y_{k+1}\tau]$ belongs to $\langle [x_m , x_n] \mid m,n \in \mathbb{Z}\rangle$ and hence $x_k\tau \in \langle v_j \mid j \in \mathbb{Z}\rangle$. By Step 5, there exists h such that $y_h\tau \notin \langle u_j \mid j \in \mathbb{Z}\rangle$, and hence $y_h\tau = \Sigma_j\ \lambda_j u_j + \mu_1 v_{t_1} + \dots + \mu_r v_{t_r}$, where $\beta_m\beta_n^{-1} = \mu_m\mu_n^{-1}$ for all m, n. It follows that $x_j\tau = \varepsilon_1 v_{t_1} + \dots + \varepsilon_r v_{t_r}$, where $\beta_m\beta_n^{-1} = \varepsilon_m\varepsilon_n^{-1}$ for all m, n, so that $[x_j\tau , x_k\tau] = 1$ for all j, k, which is impossible. Therefore $x_i\tau \in \langle u_j \mid j \in \mathbb{Z}\rangle$, and the claim follows.

(Step 7) For all $i,h \in \mathbb{Z}$, $x_i\tau = \alpha_j u_j$ and $y_h\tau = \beta_k v_k$ for some j,k.

The proof of this is similar to that of steps 1, 2, 3.

(Step 8) $x_i\tau = \alpha_i u_i$ and $y_i\tau = \gamma_i v_i$.

Let i be an even integer, and write $x_i\tau = \alpha_j u_j$; hence $\alpha_j[y_j , y_{j+1}] = \alpha_j x_j^p = (x_i\tau)^p = [y_i\tau , y_{i+1}\tau]$, so that $\{y_i\tau , y_{i+1}\tau\} \subseteq <v_j> \cup <v_{j+1}>$. Then $\alpha_j[u_j , x_{i+1}\tau] \in <y_j^p> \cup <v_{j+1}^p>$ and hence j is even and $y_i\tau \in <v_j>$; it follows also that $y_{i+1}\tau \in <v_{j+1}>$. Since $\alpha_j[u_j , x_{i+1}\tau] \in <[x_j , x_{j+1}]>$, we obtain that $x_{i+1}\tau \in <u_{j+1}>$. A simple argument now shows that $y_{i+2}\tau \in <v_{j+2}>$ and hence $x_{i+2}\tau \in <u_{j+2}>$. Therefore $x_i\tau = \alpha_i u_{i+c}$ and $y_i\tau = \gamma_i v_{i+c}$, where $c = j - i$ is a fixed integer. Since $[x_0 , x_4] = 1$, we get $c = 0$, and the claim follows.

(Step 9) $x_i\tau = \alpha u_i$ and $y_i\tau = \alpha^2 v_i$ for some fixed element $\alpha \in GF(p)$ with $\alpha^3 = 1$.

From Step 8 we obtain the equalities $\alpha_i = \gamma_i \gamma_{i+1}$, $\gamma_{2i} = \alpha_{2i} \alpha_{2i+1}$, $\gamma_{2i+1} = \alpha_{2i-1} \alpha_{2i+2}$ for all i, so that $\gamma_{2i+1} = \alpha_{2i+1}^{-1}$ and $\alpha_{2i+1} \alpha_{2i-1} \alpha_{2i+2} = 1$ for all i. Since $[x_{j-5} , x_j] = [x_{j-5} , x_{j+1}]$ for all j, it follows that $\alpha_j = \alpha_{j+1}$ and so $x_i\tau = \alpha u_i$, where $\alpha^3 = 1$, and $\gamma_{2i+1} = \alpha^{-1} = \alpha^2$, $\gamma_{2i} = \alpha_{2i} \alpha_{2i+1} = \alpha^2$.

(Step 10) $\sigma = \varphi^n\xi$ for some $\xi \in Aut_cG$.

Since $\alpha = \omega^n$ for some n, the endomorphisms φ^n and σ act in the same way on Q, and hence $\xi = \varphi^{-n}\sigma \in Aut_cG$.

(Step 11) G is an E-group.

For each $a \in G$ we have $a\sigma = (a\varphi^n)c$ for some $c \in C$ and hence $[a,a\sigma] = [a,a\varphi^n] = 1$.

We now construct two examples of infinite exponent. The first of these is obtained from 3.1 via some general homological considerations. The second one could be derived in a similar way from 3.2, but we have preferred in this case to give a more concrete, although less general, construction of it.

3.3. Let G be the group constructed in 3.1. There is an exact sequence

$$G' \otimes G_{ab} \xrightarrow{\gamma} M(G) \xrightarrow{\lambda} G_{ab} \wedge G_{ab} \xrightarrow{\delta} G'$$

where γ is the Ganea map and δ is the commutator map ([7], p. 105).

Since δ is not injective, the group $M(G)/Im\gamma$ is a non-trivial elementary abelian p-group. Let D be a p^∞-group, regarded as a trivial G-module, $f: M(G) \to D$ be a homomorphism such that $Im\gamma \leq kerf < M(G)$. We have $H^2(G,D) \cong Hom(M(G),D)$ by the Universal Coefficients Theorem, and so f defines a central extension

$$D \overset{\mu}{\rightarrowtail} H \overset{\varepsilon}{\twoheadrightarrow} G$$

Consider the homomorphism

$$\gamma ':(x\varepsilon) \otimes ((y\varepsilon)G') \in G' \otimes G_{ab} \to [x,y]\mu^{-1} \in D$$

Then $\gamma ' = \gamma f = 0$ and so $Z(H/D) = Z(H)/D$. If τ is an endomorphism of H such that $H\tau \nsubseteq Z(H)$, then by 3.1 we obtain that the endomorphism induced by τ on G is a central automorphism, so that τ acts trivially on H/Z(H) and H is a non-reduced E-group with $AutH = Aut_cH$.

3.4. Let p be a prime, $p \equiv 1 \pmod 3$, and let K be the group obtained by deleting the relation $[x_0,x_5] = [x_0,x_6]$ in the presentation of the group G of 3.2.

Let σ be an endomorphism of K. As in 3.2, and using the same notation, it can be proved that $x_i\tau = \alpha_i u_i$, $y_i\tau = \gamma_i v_i$ for all i, and

$$\alpha_i = \gamma_i\gamma_{i+1},\ \gamma_{2i} = \alpha_{2i}\alpha_{2i+1},\ \gamma_{2i+1} = \alpha_{2i-1}\alpha_{2i+2},\ \alpha_{2i+1}\alpha_{2i-1}\alpha_{2i+2} = 1;$$

it follows that $\alpha_i = \alpha_{i+1}$ if $i \neq 5$ and hence $\alpha_5 = \alpha_7^{-2} = \alpha_7$, so that $\alpha_i = \alpha_j$ for all i,j. We obtain that $[a,a\tau] = 1$ for all $a \in K$, and K is an E-group.

Write $w = [x_0, x_5x_6^{-1}]$, and denote by H the central product of K and

$$D = \langle a_n \mid n \in \mathbb{N}_0,\ a_{n+1}^p = a_n \text{ for all } n \in \mathbb{N}_0,\ a_0 = 1\rangle \cong Z(p^\infty),$$

obtained by identifying w and a_1. Then $G = H/D \cong K/\langle u\rangle$ is the group of 3.2, and hence $Z(H/D) = Z(H)/D$.

If ω is a primitive third root of 1 in GF(p), the positions

$$x_i\varphi = x_i^{\omega},\quad y_i\varphi = y_i^{\omega^2},\quad a_n\varphi = a_n^{\omega^2},$$

for $i \in \mathbb{Z}$, $n \in \mathbb{N}_0$, define a non-central automorphism of H such that $[h,h\varphi] = 1$ for all $h \in H$. Let ψ be any endomorphism of H, and denote by ψ the endomorphism induced by ψ on G; if $H\psi \not\subseteq Z(H)$, then by 3.2 we have $\psi = \varphi^i\sigma$, where σ is a central automorphism of G. Then for each $h \in H$ we get $h\psi = h\varphi^i c$ for some $c \in Z(H)$, and so $[h,h\psi] = 1$. Therefore H is a non-reduced E-group with $|\mathrm{Aut}H/\mathrm{Aut}_cH| = 3$.

References

[1] A. Caranti, Finite p-groups of exponent p^2 in which each element commutes with its endomorphic images, J. Algebra **97** (1985), 1-13.

[2] T.A. Fournelle, Automorphisms of nilpotent groups of class two with small rank, J. Austral. Math. Soc. (Ser. A) **39** (1985), 121-131.

[3] J.J. Malone, More on groups in which each element commutes with its endomorphic images, Proc. Amer. Math. Soc. **65** (1977), 209-214.

[4] D.J.S. Robinson, A property of the lower central series of a group, Math. Z. **107** (1968), 225-231.

[5] D.J.S. Robinson, *Finiteness Conditions and Generalized Soluble Groups*, Springer, Berlin (1972).

[6] D.J.S. Robinson, *A Course in the Theory of Groups*, Springer, Berlin (1982).

[7] U. Stammbach, *Homology in Group Theory*, Lecture Notes in Mathematics **359**, Springer, Berlin (1973).

POLYNOMIAL IDENTITIES WITH INVOLUTION AND THE HYPEROCTAHEDRAL GROUP.

A. Giambruno
Dipartimento di Matematica
Università di Palermo
Via Archirafi 34, 90123 Palermo

The representation theory of the symmetric group has been a useful tool in the last decade to study the polynomial identities of an algebra R over a field F of characteristic zero (or equivalently the T-ideals of the free algebra over F) (see [1],[4], [5]).

Recently a new method has been introduced that allows to study the *-polynomial identities of an algebra R with involution *, through the representation theory of the hyperoctahedral group. Here we want to illustrate such method.

Let F be a field of characteristic zero and $F\{X,*\} = F\{x_1,x_1^*,x_2,x_2^*,\ldots\}$ the free algebra with involution * in a countable set of unknowns.

<u>Def.</u> A *-T-ideal of $F\{X,*\}$ is an ideal invariant under all endomorphisms of the free algebra that commute with the involution.

<u>Def.</u> If R is an algebra with involution *, a *-polynomial identity for R is a polynomial $0 \neq f(x_1,x_1^*,\ldots,x_n,x_n^*) \in F\{X,*\}$ such that $f(a_1,a_1^*,\ldots,a_n,a_n^*) = 0$, for all $a_1,\ldots,a_n \in R$.

The *-T-ideals are strictly related to the *-polynomial identities satisfied by an algebra R. In fact, if we set

$$T(R) = \{ f(x_1,x_1^*,\ldots,x_n,x_n^*) \in F\{X,*\} \mid f \text{ is a } *\text{-polynomial identity of } R\},$$

then $T(R)$ is clearly a *-T-ideal of $F\{X,*\}$.

<u>Def.</u> A *-polynomial $f(x_1,x_1^*,\ldots,x_n,x_n^*)$ is multilinear if in every monomial of f, x_i or x_i^* $(i = 1,\ldots,n)$ appears exactly once.

Let now H_n be the hyperoctahedral group of degree n. Recall that if $Z_2 = \{1,*\}$ is the multiplicative group of order 2 and S_n is the symmetric group of degree n, then $H_n \cong Z_2^n \rtimes S_n$ and we write

$$H_n = \{ (a_1,\ldots,a_n;\sigma) \mid a_i \in Z_2,\ \sigma \in S_n \}$$

with multiplication defined by

$$(a_1,\ldots,a_n;\sigma)(b_1,\ldots,b_n;\tau) = (a_1 b_{\sigma^{-1}(1)},\ldots,a_n b_{\sigma^{-1}(n)};\sigma\tau).$$

Now,

$$V_n(*) = \mathrm{Span}_F\{x_{\sigma(1)}^{a_1}\cdots x_{\sigma(n)}^{a_n} \mid (a_1,\ldots,a_n;\sigma) \in H_n\}$$

is the space of multilinear *-polynomials in $x_1,x_1^*,\ldots,x_n,x_n^*$ and the map

$$\sum \alpha_{(a,\sigma)} x_{\sigma(1)}^{a_{\sigma^{-1}(1)}}\cdots x_{\sigma(n)}^{a_{\sigma^{-1}(n)}} \to \sum \alpha_{(a,\sigma)} (a_1,\ldots,a_n;\sigma)$$

defines an F-isomorphism of $V_n(*)$ onto FH_n.

Let now T be a *-T-ideal of $F\{X,*\}$. If $T_n = T \cap V_n(*)$ then, under the above identification, T_n becomes a left ideal of FH_n and we study the sequence of left ideals $\{T_n\}_{n\geq 1}$. Actually it has proved more convenient to study the sequence of left H_n-modules $\{V_n(*)/T_n\}_{n\geq 1}$. Let us denote by $\chi_n(T,*)$ the H_n-character of $V_n(*)/T_n$ and let us call $\{\chi_n(T,*)\}_{n\geq 1}$ the sequence of *-cocharacters of T. Since every character $\chi_n(T;*)$ is a sum of irreducible characters of H_n, the problem of determining $\chi_n(T,*)$ is reduced to that of computing the multiplicities of each irreducible character of H_n in such decomposition.

In characteristic zero it is known that there exists a one-to-one correspondence between non-equivalent irreducible representations of H_n and pairs of partitions (λ, μ) where λ is a partition of k, μ is a partition of n-k and k = 0,...,n. We write briefly $|\lambda| + |\mu| = n$. So, let us denote by $\chi_{\lambda,\mu}$ the irreducible character of H_n associated to the pair (λ, μ).

If T is a *-T-ideal we write

$$\chi_n(T,*) = \sum_{|\lambda|+|\mu|=n} m_{\lambda,\mu}\chi_{\lambda,\mu}$$

where $m_{\lambda,\mu}$ is the multiplicity of $\chi_{\lambda,\mu}$ in the given decomposition.

We state below a theorem which characterizes algebras satisfying a special kind of identity.

Let

$$d_{t+1}(x_1,\dots,x_{t+1};y_1,\dots,y_t) = \sum_{\sigma\in S_{t+1}} (\mathrm{sgn}\,\sigma)x_{\sigma(1)}y_1\dots x_{\sigma(t)}y_t x_{\sigma(t+1)}$$

be the Capelli polynomial of degree 2t+1.

If $\lambda = (\lambda_1,\dots,\lambda_r)$, $\lambda_1 \geq \lambda_2 \geq \dots \geq \lambda_r > 0$ is a partition of n, we call $r = h(\lambda)$ the height of λ ($h(\lambda)$ is the height of the corresponding Young diagram). We have

<u>Theorem 1</u> ([3,Theorem 5.8]). Let T be a *-T-ideal. Then

1) $d_{r+1}(x_1+x_1^*,\dots,x_{r+1}+x_{r+1}^*;y_1,\dots,y_r) \in T$ if and only if

$$\chi_n(T,*) = \sum_{\substack{|\lambda|+|\mu|=n \\ h(\lambda)\leq r}} m_{\lambda,\mu}\chi_{\lambda,\mu}.$$

2) $d_{u+1}(x_1-x_1^*,\dots,x_{u+1}-x_{u+1}^*;y_1,\dots,y_u) \in T$ if and only if

$$\chi_n(T,*) = \sum_{\substack{|\lambda|+|\mu|=n \\ h(\mu)\leq u}} m_{\lambda,\mu}\chi_{\lambda,\mu}.$$

This theorem has an application to $M_k(F)$, the algebra of kxk matrices over F.

Suppose F is an algebraically closed field of characteristic zero. In this case $M_k(F)$ has only two possible involutions:

1) the transpose involution (* = t): if $A \in M_k(F)$, $A^* = A^t$, where A^t is the usual transpose of A;

2) the symplectic involution (* = s): in this case k is even and * is given by $(A_{ij})^* = (A^*_{ij})$ where the A_{ij}'s are 2x2 matrices over F with involution given by

$$\begin{pmatrix} a & b \\ c & d \end{pmatrix}^* = \begin{pmatrix} d & -b \\ -c & a \end{pmatrix}$$

If T is the *-T-ideal of identities of $M_k(F)$ (* = t or s), we write $\chi_n(T,*) = \chi_n(M_k(F),*)$. Also let $r = 1/2k(k+1)$ and $u = 1/2k(k-1)$. We have

<u>Corollary.</u> Let F be an algebraically closed field of characteristic zero. Then

$$\chi_n(M_k(F),t) = \sum_{\substack{|\lambda|+|\mu|=n \\ h(\lambda) \leq r \\ h(\mu) \leq u}} m_{\lambda,\mu}\chi_{\lambda,\mu}.$$

and

$$\chi_n(M_k(F),s) = \sum_{\substack{|\lambda|+|\mu|=n \\ h(\lambda) \leq u \\ h(\mu) \leq r}} m_{\lambda,\mu}\chi_{\lambda,\mu}.$$

Moreover, if * = t (* = s, respectively), there exist n and partitions λ, μ, $|\lambda|+|\mu| = n$ such that $h(\lambda) = r$ ($h(\lambda) = u$), $h(\mu) = u$ ($h(\mu) = r$) and $m_{\lambda,\mu} = m_{\lambda,\mu}(M_k(F),*)$ is non zero.

Proof. Since char F = 2, $M_k(F) = M_k(F)^+ \oplus M_k(F)^-$ where $M_k(F)^+ = \{ A + A^* \mid A \in M_k(F) \}$ and $M_k(F)^- = \{ A - A^* \mid A \in M_k(F) \}$ are the sets of symmetric and skew elements of $M_k(F)$ respectively.

Now, an easy calculation shows that

$$\dim M_k(F)^+ = \begin{cases} r & \text{if } * = t \\ u & \text{if } * = s \end{cases}$$

and

$$\dim M_k(F)^- = \begin{cases} u & \text{if } * = t \\ r & \text{if } * = s. \end{cases}$$

Suppose $* = s$; then , by [6, Theorem 1.4.34]

$$d_{u+1}(x_1+x_1^*,\dots,x_{u+1}+x_{u+1}^*;y_1,\dots,y_u)$$

and

$$d_{r+1}(x_1-x_1^*,\dots,x_{r+1}-x_{r+1}^*;y_1,\dots,y_r)$$

are *-identities for $M_k(F)$ whereas

$$d_u(x_1+x_1^*,\dots,x_u+x_u^*;y_1,\dots,y_{u-1})$$

and

$$d_r(x_1-x_1^*,\dots,x_r-x_r^*;y_1,\dots,y_{r-1})$$

are not *-identities for $M_k(F)$. An analogue argument holds in case $* = t$.

The free algebra $F\{X,*\}$ is multigraded by degree; also if T is a *-T-ideal , T is homogeneous, that is every homogeneous component of a polynomial in T lies in T. To study the homogeneous polynomials in T, we proceed as follows: let U and V be n-dimensional vector spaces over F with bases $\{u_1,\dots,u_n\}$ and $\{v_1,\dots,v_n\}$ respectively. Let

$$W = \underbrace{(U \oplus V) \otimes \dots \otimes (U \oplus V)}_{n}.$$

Then W can be identified with the space of homogeneous degree n polynomials with involution in the variables $x_i = u_i + v_i$ and $x_i^* = u_i - v_i$.

The group $GL(U) \times GL(V)$ acts naturally on the space $U \oplus V$: if

$C = (A,B) \in GL(U) \times GL(V)$ then $C(u+v) = Au + Bv$. This in turn defines an action of $GL(U) \times GL(V)$ on W: if $C \in GL(U) \times GL(V)$ and $w_1 \otimes \dots \otimes w_n \in W$ then

$$C(w_1 \otimes \dots \otimes w_n) = Cw_1 \otimes \dots \otimes Cw_n.$$

The action of $GL(U) \times GL(V)$ on $U \oplus V$ also defines an action of H_n on $U \oplus V$ in the following way: H_n acts on U as the matrix permutations (i.e., as its quotient $H_n/Z_2^n \simeq S_n$) and H_n acts on V as the subgroup of $GL(n)$ generated by the matrix permutations and the matrices $\{diag(e_1,\dots,e_n) \mid e_i = \pm 1\} \simeq Z_2^n$.

Now, if $\alpha_1,\dots,\alpha_n \in F - \{0\}$, define $T_{\alpha_1 \dots \alpha_n} \in GL(U) \times GL(V)$ by setting

$$T_{\alpha_1 \dots \alpha_n}(u_i) = \alpha_i u_i \quad \text{and} \quad T_{\alpha_1 \dots \alpha_n}(v_i) = \alpha_i v_i,$$

for $i = 1,\dots,n$. Let

$$T = \{T_{\alpha_1 \dots \alpha_n} \mid \alpha_1,\dots,\alpha_n \in F - \{0\}\}.$$

If M is a $GL(U) \times GL(V)$ - submodule of W, define the multilinear space of M as

$$M^{mult} = \{w \in M \mid T_{\alpha_1 \dots \alpha_n} w = \alpha_1 \dots \alpha_n w, \text{ for all } T_{\alpha_1 \dots \alpha_n} \in T\}.$$

It is clear that under the above identification, $W^{mult} = V_n(*)$. Also we have

<u>Lemma.</u> If M is a $GL(U) \times GL(V)$ - submodule of W, then M^{mult} is an H_n-module.

<u>Proof.</u> Let $w \in M^{mult}$ and $g = (a_1,\dots,a_n;\sigma) \in H_n$. Then, for $\alpha_1,\dots,\alpha_n \in F - \{0\}$,

$$T_{\alpha_1 \dots \alpha_n}(gw) = gg^{-1}T_{\alpha_1 \dots \alpha_n} gw = gT_{\alpha_{\sigma(1)} \dots \alpha_{\sigma(n)}}(w) =$$

$$g\alpha_{\sigma(1)} \dots \alpha_{\sigma(n)} w = \alpha_1 \dots \alpha_n (gw).$$

This implies that $gw \in M^{mult}$.

Now, The representation theory of $GL(U) \times GL(V)$ acting on W is known. There exists a one-to-one correspondence between irreducible non-equivalent representations of $GL(U) \times GL(V)$ and pairs of partitions (λ,μ) where λ is a partition of k, μ is a partition of $n - k$ $(k = 0,\dots,n)$. So, let us write $\psi_{\lambda,\mu}$ for the irreducible character of $GL(U) \times GL(V)$ associated to the pair (λ,μ). ALso, if M is a $GL(U) \times GL(V)$ - module, let us write $\psi(M)$ for the character of M. The relation between $GL(U) \times GL(V)$ - characters and H_n-characters is given by the following

<u>Theorem 2</u> ([2, Theorem 1]). Let $M \subset W$ be a $GL(U) \times GL(V)$ - module. If

$$\psi(M) = \sum_{|\lambda|+|\mu|=n} m_{\lambda,\mu}\psi_{\lambda,\mu}$$

and

$$\chi(M^{mult}) = \sum_{|\lambda|+|\mu|=n} m'_{\lambda,\mu}\chi_{\lambda,\mu}$$

then $m_{\lambda,\mu} = m'_{\lambda,\mu}$ for all λ,μ.

This theorem has an immediate application to the space of homogeneous *-polynomials of a *-T-ideal: let T be a *-T-ideal ; then $T \cap W$ is the space of homogeneous degree n *-polynomials in $T \cap F\{x_1,x_1^*,\dots,x_n,x_n^*\}$ and $T \cap W$ is a $GL(U) \times GL(V)$ - module. We want to study the $GL(U) \times GL(V)$ module $W/T \cap W$ and we want to relate its character to the n-th cocharacter of T.

Let $\psi_n(T,*)$ be the $GL(U) \times GL(V)$ - character of $W/T \cap W$. As a corollary of the previous theorem we obtain the following

<u>Corollary.</u> Let T be a *-T-ideal of $F\{X,*\}$. If

$$\psi_n(T,*) = \sum_{|\lambda|+|\mu|=n} m_{\lambda,\mu}\psi_{\lambda,\mu}$$

and

$$\chi_n(T,*) = \sum_{|\lambda|+|\mu|=n} m'_{\lambda,\mu}\chi_{\lambda,\mu}$$

then $m_{\lambda,\mu} = m'_{\lambda,\mu}$, for all λ,μ.

REFERENCES

1. E. Formanek, The polynomial identities of matrices, in "Algebraists' Homage" (S. A. Amitsur et al.,Eds.) Contemporary Mathematics, 41-79, Amer. Math. Soc., Providence, 1982.

2. A. Giambruno, GL x GL - representations and *-polynomial identities, Comm. Algebra, (to appear).

3. A. Giambruno and A. Regev, Wreath products and PI-algebras, J. Pure Applied Algebra, 35 (1985), 133-150.

4. A. Regev, Existence of identities in $A \otimes B$, Israel J. Math., 11 (1972), 131-152.

5. A. Regev, The representation theory of S_n and explicit identities for PI-algebras, J. Algebra, 51 (1978), 25-40.

6. L. H. Rowen, "Polynomial Identities in Ring Theory", Academic Press, New York-London, 1980.

AUTOMORPHISMS OF INDUCED EXTENSIONS

Fletcher Gross*

University of Utah

Salt Lake City, 84112 Utah, U.S.A.

Suppose M is a normal subgroup of the group G and M is a direct product $R_1 \times \dots \times R_n$ where $R_1, \dots, R_n$ are all the conjugates of R_1 in G. (The model I have in mind is where M is a non-abelian minimal normal subgroup of a finite group, but we don't need to be that specific: Indeed, most of the results I will describe remain true without any finiteness assumptions whatsoever. Although if R_1 has infinitely many conjugates, then M must be the unrestricted direct product). Let $R = R_1$, $N = N_G(R)$, and $K = R_2 \times \dots \times R_n$. It is shown in [2] that G can be constructed from knowledge only of the groups N/K and G/M. Our interest now is to show that the automorphisms of G are determined by their actions on G/M and N/K.

Suppose σ is an automorphism of G. Then σ induces an isomorphism α of G/M onto G/M^σ and an isomorphism β of N/K onto N^σ/K^σ. Further, α and β both induce the same isomorphism of N/M onto N^σ/M^σ. The basic result deals with the converse: If α and β are isomorphisms defined on G/M and N/K, respectively, and if α and β both induce the same isomorphism of N/M, is there an automorphism σ of G which induces both α and β? If we impose certain mild conditions, the answer is yes.

To be more specific, consider for example, the group A consisting of all automorphisms of G which map R onto itself. Then $A \cap \mathrm{Inn}(G)$ is $\mathrm{Inn}_N(G)$, the automorphisms of G obtained by conjugating by elements of N, every element of A fixes K, and AInn(G) is the group of all automorphisms σ such that R and R^σ are conjugate in G. Now let B be the set of all ordered pairs (α,β) satisfying the following:

α is an automorphism of G/M, β is an automorphism of N/K,

---

* The author acknowledges the kind hospitality of the Mathematics Institute of the University of Florence and C.N.R. support during the course of this work.

α fixes N/M, β fixes M/K, and α and β induce the same automorphism of N/M. Then B is a subgroup of $\mathrm{Aut}(G/M) \times \mathrm{Aut}(N/K)$. The relationship between A and B is given by

THEOREM 1. $B \cong A/\mathrm{Inn}_K(G)$.

Proof. Here, we only prove this for G a finite group. First note that there is an obvious homomorphism f of A into B where if $\gamma \in A$, then f maps γ into the ordered pair of automorphisms induced by γ. If $\gamma \in \mathrm{Kernel}(f)$, then γ must fix each element of G/M and N/K. It follows from Corollary 3.2 and Theorem 4.1 (1) of [2] that $\gamma \in \mathrm{Inn}_K(G)$. Hence, the kernel of f is $\mathrm{Inn}_K(G)$ and the theorem will follow once we show that f maps A onto B.

Assume then that $(\alpha,\beta) \in B$. Let σ be the natural homomorphism of G onto G/M, let e be the natural homomorphism of N onto N/K, and let μ be the homomorphism of N/K into G/M given by $\mu(Kn) = Mn$ for all $n \in N$. Then $\mu e = \sigma_N$, the restriction of σ to N, and, since α and β must induce the same automorphism of N/M, the following diagram commutes.

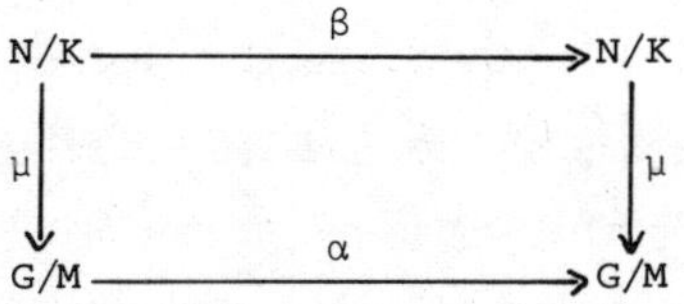

Now let $\tau = \alpha\sigma$ and $\rho = \beta e$. Then τ is a homomorphism of G onto G/M and

$$\{X \in G \mid \tau(X) \in N/M\} = N \quad .$$

Also, ρ is a homomorphism of N onto N/K and

$$\mu\rho = (\mu\beta)e = \alpha(\mu e) = \alpha\sigma_N = \tau_N \quad .$$

Theorem 4.1 (2) of [2] now implies that there is a homomorphism γ of G into G such that $\tau = \sigma\gamma$ and $\rho = e\gamma_N$. Let $L = \mathrm{Kernel}(\gamma)$. Then

$$1 = \sigma\gamma(L) = \tau(L)$$

and so $L \leq \mathrm{Kernel}(\tau) = M \leq N$. Then L is contained in the domain of ρ and thus

$$\rho(L) = e\gamma(L) = 1 \quad .$$

Hence, $L \leq \mathrm{Kernel}(\rho)$. Since β is an automorphism, the kernel of ρ is the same as the kernel of e which is K. Thus $L \leq K$. Since L is a normal subgroup of G and since the intersection of all conjugates of K is 1, we see that $L = 1$. Hence γ maps G isomorphically into G. We are assuming that G is finite and so it follows that γ is an automorphism of G. Now

$$M = \mathrm{Kernel}(\tau) = \mathrm{Kernel}(\sigma\gamma) = \gamma^{-1}(M) \quad .$$

This implies both that $\gamma(M) = M$ and $\gamma^{-1}(K) \leq N$. Then

$$K = \mathrm{Kernel}(\rho) = \mathrm{Kernel}(e\gamma_N) = \gamma^{-1}(K)$$

and so $\gamma(K) = K$. Therefore, $\gamma \in A$. If $g \in G$ and $n \in N$, then

$$\alpha(Mg) = \alpha\sigma(g) = \tau(g) = \sigma\gamma(g) = M\gamma(g)$$

and

$$\beta(Kn) = \beta e(n) = \rho(n) = e\gamma_N(n) = K\gamma(n) \quad .$$

It now follows that both α and β are induced by γ. Hence $f(\gamma) = (\alpha,\beta)$. This proves the theorem.

Often, it is not difficult to compute B and so Theorem 1 can be used to obtain information about A. To illustrate how this works, suppose G is the permutational wreath product R Wr (P,I) where P is a transitive permutation group acting on the set I. Let M be the base subgroup R^I. Then M is the direct product of $|I|$ copies of R and P acts on M by permuting the copies of R. Let H be the stabilizer in P of the point $i_o \in I$. Let $K = \{m \in M | m(i_o) = 1\}$ and $N = HM$. Then the hypothesis of Theorem 1 is satisfied.

In the situation just described, $N/K \cong R \times H$ and $G/M \cong P$. If $(\alpha,\beta) \in B$, then α is an automorphism of P which fixes H. Hence $\alpha \in N_{\mathrm{Aut}(P)}(H)$. Now β is an automorphism of $R \times H$ which fixes R. The automorphisms of $R \times H$ which fix R form a group isomorphic to the semi-direct product of $\mathrm{Hom}(H,Z(R))$, an abelian group, by $(\mathrm{Aut}(R) \times \mathrm{Aut}(H))$. (Here if $\lambda \in \mathrm{Hom}(H,Z(R))$, then λ determines the automorphism $(r,h) \to (r\lambda(h),h)$ of $R \times H$). Now β must agree with α on H. Thus we may pick α arbitrarily from $N_{\mathrm{Aut}(P)}(H)$ and then β may be any element of

$$\mathrm{Hom}(H,Z(R))(\mathrm{Aut}(R) \times \{\alpha_H\}) \quad .$$

It now follows that $B \cong \mathrm{Hom}(H,Z(R))(\mathrm{Aut}(R) \times N_{\mathrm{Aut}(P)}(H))$. The homomorphism mapping A onto B maps $\mathrm{Inn}_N(G)$ onto $\mathrm{Inn}(R) \times \mathrm{Inn}_H(P)$. Since $A \cap \mathrm{Inn}(G) = \mathrm{Inn}_N(G)$, we have proved the following.

Corollary. With the notation as above,

$$A/\mathrm{Inn}_K(G) \cong \mathrm{Hom}(H,Z(R))(\mathrm{Aut}(R) \times N_{\mathrm{Aut}(P)}(H)) \quad ,$$

and

$$\mathrm{AInn}(G)/\mathrm{Inn}(G) \cong \mathrm{Hom}(H,Z(R))(\mathrm{Aut}(R) \times (N_{\mathrm{Aut}(P)}(H)/\mathrm{Inn}_H(P))) \quad .$$

As an aside, if Q is the normalizer of P in the group of all permutations of I, then $N_{\mathrm{Aut}(P)}(H)/\mathrm{Inn}_H(P) \cong Q/P$, a fact that sometimes simplifies computations.

In many cases, AInn(G) = Aut(G); if this is not the case then to determine the full automorphism group, we need to find the orbit of R under Aut(G). In some cases, this has been done and typical of the results obtained are the final 2 theorems. In order not to get too technical, I have only included examples involving wreath products and I have just given order formulas although in the last theorem I have completely determined the automorphisms.

THEOREM 2. Let G be the standard wreath product R Wr P with R and P finite groups which are not both of order 2. Then

$$|\mathrm{Aut}(G)| \le |\mathrm{Aut}(R)| \; |\mathrm{Aut}(P)| \; [|R| \; |\mathrm{Hom}(R,R)|]^{|P|-1} \quad .$$

If, in addition, R and P are both p-groups for any prime p, and R is abelian, then (we still are assuming that R and P do not both have order 2) the above upper bound is attained, i.e., in this case

$$|\mathrm{Aut}(G)| = |\mathrm{Aut}(R)| \; |\mathrm{Aut}(P)| \; [|R| \; |\mathrm{Hom}(R,R)|]^{|P|-1} \quad .$$

THEOREM 3. Let P be a transitive permutation group on the finite set I, $|I| > 1$, let H be the stabilizer of a point, and let m be the number of orbits H has on I. Let R be a finite, non-abelian, indecomposable group and let G = R Wr (P,I). Let A be the group of all automorphisms of G which fix (not necessarily pointwise) the base subgroup R^I. Then $\mathrm{Inn}(G) \le A \le \mathrm{Aut}(G)$ and

$$|A/\mathrm{Inn}(G)| = |\mathrm{Hom}(H,Z(R)| \; |\mathrm{Out}(R)| \; |N_{\mathrm{Aut}(P)}(H)/\mathrm{Inn}_H(P)| \; |\mathrm{Hom}(R,Z(R)|^{m-1} \quad .$$

Further, A = Aut(G) if any of the following hold:

(1) P is regular on I (i.e., H = 1) and either $|I| > 2$ or R is not a special dihedral group (see [1]).

(2) P is primitive on I and $|I| > 2$.

(3) H is contained in some core-free maximal subgroup of G and $|I| > 2$.

(4) $O_2(P) = 1$.

(5) $O^2(R) = R$.

The proofs of Theorems 2 and 3 will appear in [3] and[4].

REFERENCES

[1] P.M. Neumann, On the structure of standard wreath products of groups, Math. Z. 84 (1964), 343-373.

[2] F. Gross and L.G. Kovacs, On normal subgroups which are direct products, J. Algebra 90 (1984), 133-168.

[3] F. Gross, Automorphisms of permutational wreath products, J. Algebra, to appear.

[4] F. Gross, Automorphisms of induced extensions, J. Algebra, to appear.

On dimension subgroups relative to certain product ideals

C.K.Gupta, N.D.Gupta and F.Levin

1. Introduction

Let **f** denote the augmentation ideal of the integral group ring **ZF** of a free group F and **x** be any ideal of **ZF** contained in **f**. For each $n \geq 1$, **x** induces a normal subgroup D(n,**x**) of F defined by D(n,**x**) = $F \cap (1+\mathbf{x}+\mathbf{f}^n)$, which we call the n-th dimension subgroup of F relative to the ideal **x**. A classical result due to Magnus [3] states that D(n,**0**) = $\gamma_n(F)$, the n-th term of the lower central series of F. For any normal subgroup $R \leq F$, the identification of D(n,**r**) with **r** = **ZF**(R-1) is the well- known dimension subgroup problem (we refer to Passi [5] for more detailed background information on this problem).It had long been conjectured that the dimension subgroup D(n,**r**) coincides with R $\gamma_n(F)$ for all n and all R. However, Rips [6] has constructed a counter-example to this conjecture by constructing a finite 2-group F/R such that D(4,**r**) $\neq$ R $\gamma_4(F)$. This in turn yields that, for Rips' group F/R, D(4,**fr**) $\nleq$ R $\gamma_4(F)$ (see , for instance, [1] Lemma 4) . On the other hand, if F/R is a free abelian group then Gupta [1] has shown that a stronger result holds, namely, D(n,**fr**) =R'$\gamma_n(F)$ for all $n \geq 1$, where R' = $\gamma_2(R)$ is the commutator subgroup of R. More generally,if F/R is a finitely generated abelian group then Gupta, Hales and Passi [2] have shown that there exists $n_0 = n_0(F/R)$ such that D(n,**fr**) = R'γ_n (F) for all $n \geq n_0$. These results have additional important implications towards the solution of the dimension subgroup problem for metabelian groups. It is, therefore, natural to ask if the corresponding results hold when F/R is a finitely generated nilpotent group. In this paper we settle this problem when F/R is a free nilpotent group and prove that, in this case, D(n,**fr**) = R'$\gamma_n(F)$ for all $n \geq 1$.

2.Basic commutators mod $[\gamma_{c+1} (F), \gamma_{c+1} (F)]$

For $n \geq 2c$, $c \geq 2$, we shall need a detailed analysis of the structure of a basic commutator of weight n which lies in $[\gamma_c(F), \gamma_c(F)]$ but not in $[\gamma_{c+1}(F), \gamma_{c+1}(F)]$. The following details can be found in Chapter 5 of Magnus, Karrass and Solitar [4] which is also our source for any unexplained notation used in the sequel.

For $n \geq 2c$, $c \geq 2$, let $t^*= (t(1),...,t(c))$ be a c-tuple of non-negative integers satisfying

$$t(1) +2\, t(2) +...+c\, t(c) = n\ , \quad t(1) \geq 2,\ \ t(c) \geq 1. \tag{1}$$

Let $c_{t,k}$ be a commutator of weight n defined by

$$c_{\underline{t},k} = [\, b_{1,1,k}, \ldots, b_{1,t(1),k}, \ldots, b_{c,1,k}, \ldots, b_{c,t(c),k}] \ , \tag{2}$$

such that $\underline{t}$ satisfies (1) and the following three conditions hold:

(i) each $b_{i,j,k}$ is a basic commutator of weight i ;

(ii) $b_{1,1,k} > b_{1,2,k} \leq b_{1,3,k} \leq \ldots \leq b_{2,1,k} \leq \ldots \leq b_{2,t(2),k} \leq \ldots \leq b_{c,1,k} \leq \ldots \leq b_{c,t(c),k}$, with respect to the standard ordering of the basic commutators;

(iii) if, for each $1 \leq j \leq c$, $1 \leq m \leq t(j)$, $c_{\underline{t},j,m,k}$ is an initial sub-commutator of $c_{\underline{t},k}$ defined by

$$c_{\underline{t},j,m,k} = [\, b_{1,1,k}, \ldots, b_{1,t(1),k}, \ldots, b_{j,1,k}, \ldots, b_{j,m,k}] \ , \tag{3}$$

then $c_{\underline{t},j,m,k}$ is a basic commutator with $c_{\underline{t},j,m,k} > b_{j,m+1,k}$ if $m < t(j)$ and $c_{\underline{t},j,m,k} > b_{j+1,1,k}$ if $m=t(j)$, $j < c$.

It is easy to see that $c_{\underline{t},k}$ characterizes all basic commutators of weight n in $[\gamma_c(F), \gamma_c(F)]$ which are not in $[\gamma_{c+1}(F), \gamma_{c+1}(F)]$. Thus we have the following lemma.

Lemma 2.1. *Let* $w \in (\gamma_n(F) \cap [\gamma_c(F), \gamma_c(F)])\, \gamma_{n+1}(F)$, $n \geq 2c$, $c \geq 2$. *Then modulo* $[\gamma_{c+1}(F), \gamma_{c+1}(F)]\, \gamma_{n+1}(F)$,

$$w \equiv \Pi_{\underline{t}} \Pi_k (c_{\underline{t},k})^{a(\underline{t},k)} \quad , a(\underline{t}, k) \in \mathbf{Z},$$

where the product is taken over all m*-tuples* $\underline{t}$ *satisfying* (1) *and the* $c_{\underline{t},k}$ *are as in* (2).

3. Basic products mod $\mathbf{a}_{c+1}$

For $c \geq 1$, let $\mathbf{a}_c$ denote the ideal $\mathbf{Z}F(\gamma_c(F)-1)$ of $\mathbf{Z}F$. A basic product mod $\mathbf{a}_{c+1}$ of length n is an element of $\mathbf{f}^n$ of the form

$$\beta = (b_1-1)\ldots(b_t-1) \tag{4}$$

satisfying

(i) $b_1 \leq \ldots \leq b_t$, and

(ii) $wt(b_1) + \ldots + wt(b_t) = n$ ($wt(b_i)$ = weight of b_i).

It follows from the Hall Basis Theorem ([4] , Theorem 5.8) that the products mod $\mathbf{a}_{c+1}$ of length n form a basis for $\mathbf{f}^n$ modulo $(\mathbf{a}_{c+1} + \mathbf{f}^{n+1})$. We define a total ordering of the above basic products as follows.

If $\beta = (b_1-1) \ldots (b_t-1)$ and $\beta' = (b_1'-1) \cdots (b_s'-1)$ are basic products mod $\mathbf{a}_{c+1}$ of length n then we define $\beta < \beta'$ if either

(i) $(wt(b_t), \ldots, wt(b_1)) < (wt(b_s'), \ldots, wt(b_1'))$

via the usual lexicographic ordering , Or

(ii) $s = t$ and $(b_t, \ldots, b_1) < (b_t', \ldots, b_1')$

via the lexicographic ordering of the basic commutators.

The following lemma is a straight forward consequence of the above ordering of basic products.

Lemma 3.1 *Let* β *be as in* (4). *For any* $1 \leq m < t$, *let* $\beta_m = (b_{m+1}-1) \ldots (b_t-1)$, $n^* = wt(b_1) + \ldots + wt(b_m)$ *and* $q = wt(b_{m+1})$.*Then for any* $\alpha \in \mathbf{f}^{n^*} \cap \mathbf{a}_{q+1}$, $\alpha\beta_m$ *is a* **Z**-*linear* *combination of basic products* mod $\mathbf{a}_{c+1}$ *of length* n *each of which is greater than* β.

4.Dimension subgroups relative to $\mathbf{f}\,\mathbf{a}_c$

In this section we prove our main result which can be stated as follows.

Theorem 4.1. *For all* $n \geq 1,\ c \geq 2,\ F \cap (1 + \mathbf{fa}_c + \mathbf{f}^{n+1}) = [\gamma_c(F), \gamma_c(F)]\, \gamma_{n+1}(F)$.

Proof. Since $[\gamma_c(F), \gamma_c(F)]\, \gamma_{n+1}(F) \leq F \cap (1 + \mathbf{fa}_c + \mathbf{f}^{n+1})$, we need only prove the reverse inequality. The proof is by induction on $c \geq 2$. For $c = 2$ the result is Theorem B of Gupta [1]. Thus suppose that $c \geq 3$ and let $w \in F \cap (1 + \mathbf{fa}_{c+1} + \mathbf{f}^{n+1})$. By the induction hypothesis, $w \in [\gamma_c(F), \gamma_c(F)]\, \gamma_{n+1}(F)$. If $n + 1 < 2c+1$ then

$$w \in \gamma_{n+1}(F) = [\gamma_{c+1}(F), \gamma_{c+1}(F)]\gamma_{n+1}(F),$$

as required. Hence we may assume, without loss of generality, that $n \geq 2c$ and further, by a simple induction on n, that $w \in \gamma_n(F) \cap [\gamma_c(F), \gamma_c(F)]$. With respect to any free generator x of F we may write w additively as

$$w = w_0 + w_1 + \ldots + w_{n-1}, \tag{5}$$

where w_j denotes the component whose commutator factors involve precisely j occurrences of x. For any prime p, replacing x by x^p in (5) yields the congruence

$$w_0 + pw_1 + \ldots + p^{n-1}w_{n-1} \equiv 0 \tag{6}$$

mod $K \cap [\gamma_c(F), \gamma_c(F)] \cap \gamma_n(F)$, where $K = F \cap (1 + \mathbf{fa}_{c+1} + \mathbf{f}^{n+1})$. Letting p range over a set $\{p_1, \ldots, p_n\}$ of n distinct primes yields a system of n congruences of the form (6) whose coefficient matrix is the non-singular Van der Monde matrix

$$\begin{bmatrix} 1 & p_1 & \ldots p_1^{n-1} \\ 1 & p_2 & \ldots p_2^{n-1} \\ \cdot & \cdot & \cdots \\ 1 & p_n & \ldots p_n^{n-1} \end{bmatrix}.$$

It follows that each $w_j \equiv 0 \mod K \cap [\gamma_c(F), \gamma_c(F)] \cap \gamma_n(F)$. Thus we may assume, without loss of generality, that w itself is a homogeneous element of $K \cap [\gamma_c(F), \gamma_c(F)] \cap \gamma_n(F)$ and prove that $w \in [\gamma_{c+1}(F), \gamma_{c+1}(F)]\, \gamma_{n+1}(F)$.

Let x be a free generator of F which occurs j times, $j \geq 2$, in each factor of w. Replacing x by $x_1 \ldots x_j$ and expanding, using linearity, yields a component w^* whose factors involve each of $x_1, \ldots, x_j$ and w^* itself lies in $K \cap [\gamma_c(F), \gamma_c(F)] \cap \gamma_n(F)$. Suppose we can now prove that $w^* \in [\gamma_{c+1}(F), \gamma_{c+1}(F)]\gamma_{n+1}(F)$. Then setting $x_1 = \ldots = x_j = x$ will yield $w^{j!} \in [\gamma_{c+1}(F), \gamma_{c+1}(F)]\, \gamma_{n+1}(F)$ and, in turn, $w \in [\gamma_{c+1}(F), \gamma_{c+1}(F)]\, \gamma_{n+1}(F)$ (since the quotient $F/[\gamma_{c+1}(F), \gamma_{c+1}(F)]\gamma_{n+1}(F)$ is torsion free). Thus we may further assume, without loss of generality, that $F = \langle x_1, \ldots, x_n \rangle$ and $w \in K \cap [\gamma_c(F), \gamma_c(F)] \cap \gamma_n(F)$ is a product of commutators of weight n with weight 1 in each of the generators $x_1,\ldots,x_n$. We proceed to prove that $w \in [\gamma_{c+1}(F),\gamma_{c+1},(F)]\gamma_{n+1}(F)$. By Lemma 2.1, w is congruent, mod$[\gamma_{c+1}(F),\gamma_{c+1}(F)]\gamma_{n+1}(F)$, to a product of basic commutators of the form $c_{\underline{t},k}$ as defined by (2), and by hypothesis $w - 1 \in \mathbf{f}(\mathbf{a}_{c+1}+ \mathbf{f}^n)$. We write

$$w - 1 = \Sigma_x(x-1)\,\partial_x(w)\,,\quad x \in \{x_1,\ldots,x_n\}.$$

Since $\mathbf{f}$ is a right ZF module with basis $\{x_i - 1; i = 1,\ldots,n\}$, it follows that $\partial_x(w) \in \mathbf{a}_{c+1} + \mathbf{f}^n$ for all x. The proof consists in showing that if $w \notin [\gamma_{c+1}(F),\gamma_{c+1}(F)]\gamma_{n+1}(F)$ then, for some $x \in \{x_1,\ldots,x_n\}$, $\partial_x(w) \notin \mathbf{a}_{c+1} + \mathbf{f}^n$.

Indeed, let $w \notin [\gamma_{c+1}(F),\gamma_{c+1}(F)]\gamma_{n+1}(F)$ and let

$$c_{\underline{t}} = [b_{1,1},b_{1,2}\ldots,b_{1,t(1)}\, b_{2,1},\ldots,b_{2\,,t(2)}\,,\ldots,b_{c,1},\ldots,b_{c,t(c)}]$$

be the factor of w which is minimal among all the non-trivial factors of w. Let $w = c_{\underline{t}}{}^a w'$ with $0 \neq a \in \mathbf{Z}$ such that $c_{\underline{t}}$ is not a factor of w'. Set $b_{11} = x$. Then

$$\partial_x(w) = a\,\partial_x(c_{\underline{t}}) + \partial_x(w')$$

and

$$\partial_x(c_{\underline{t}}) = (b_{1,2}-1)\ldots b_{1,t(1)}-1)\,\ldots\,(b_{c,1}-1)\,\ldots\,(b_{c,t(c)}-1)$$

is a basic product mod $\mathbf{a}_{c+1}$ of length n-1. Hence to complete the proof it suffices to show that $\partial_x(c_{\underline{t}})$ does not occur in $\partial_x(w')$ when expressed in terms of basic products mod $\mathbf{a}_{c+1} + \mathbf{f}^n$ of length n-1. Let $c_{\underline{t}'}$ be a factor of w' and suppose that $\partial_x(c_{\underline{t}})$ occurs in $\partial_x(c_{\underline{t}'})$ when expressed in terms of basic products. Then there are two cases to be considered:(i) $\underline{t}' > \underline{t}$ and (ii) $\underline{t}' = \underline{t}$.

<u>Case I</u> ($\underline{t}' > \underline{t}$). Let $c_{\underline{t}'} = [b'_{1,1},\ldots,b'_{c,t'(c)}]$ be as in (2). Since $\underline{t}' > \underline{t}$, there exists i,j such that $wt(b'_{i,j}) > wt(b_{i,j})$ and $wt(b'_{p,q}) = wt(b_{p,q})$ for all $p > i, q > j$. For $\partial_x(c_{\underline{t}}) = (b_{1,2}-1)\ldots(b_{c,t(c)}-1)$ to be a summand of $\partial_x(c_{\underline{t}'})$, $x = b_{1,1}$ must occur in $b'_{i,j}$ itself. Writing $c_{\underline{t}'} = [\alpha',b'_{i,j},\ldots,b'_{c,t'(c)}]$ we note that $\alpha' > b'_{i,j}$ and $(\alpha' - 1)$ occurs as a factor of $\partial_x(c_{\underline{t}'})$. Thus, by Lemma 3.1, $\partial_x(c_{\underline{t}'})$ is a Z-linear sum of basic products which are strictly greater than $\partial_x(c_{\underline{t}})$.

<u>Case II</u> ($\underline{t}' = \underline{t}$). In this case the weights of $b'_{i,j}$ and $b_{i,j}$ are the same for each i,j but for some i,j, $b'_{i,j} > b_{i,j}$ in the ordering of basic commutators. As before, we write $c_{\underline{t}'} = [\alpha', b'_{i,j},\ldots,b'_{c,t'(c)}]$ with x occurring in $b'_{i,j}$. Since $\alpha' > b'_{i,j}$, as in case I, if $b'_{i,j} \neq b'_{1,1}$ or $b'_{1,2}$ then $\partial_x(c_{\underline{t}'})$ is a **Z**-linear sum of basic products strictly greater than $\partial_x(c_{\underline{t}})$. Thus we must have $x = b'_{1,1}$ or $b'_{1,2}$. If $x = b'_{1,2}$ then $c_{\underline{t}'} < c_{\underline{t}}$ contrary to our choice of $c_{\underline{t}}$. If $x = b'_{1,1}$ then $c_{\underline{t}}' = c_{\underline{t}}$ contrary to the fact that $c_{\underline{t}}$ is not a factor of w'. This completes the proof of our main theorem.

References

[1] Narain Gupta, On the dimension subgroups of metabelian groups, J. Pure Appl. Algebra 24(1982), 1-6.

[2] N. D. Gupta, A. H. Hales and I. B. S. Passi, Dimension subgroups of metabelian groups,J. reine u.angew. Math. 346(1983), 194-198.

[3] Wilhelm Magnus,Über Beziehungen zwischen höheren Kommutatoren, J.reine u.angew. Math. 177(1937), 105-115.

[4] W. Magnus, A. Karrass and D. Solitar, Combinatorial GroupTheory, Interscience (1966), New York

[5] Inder Bir S. Passi, Group Rings and Their Augmentation Ideals,Springer-Verlag Lecture Notes 715(1979).

[6] E. Rips, On the fourth integer dimension subgroup, Israel J. Math. 12(1972), 342-346.

University of Manitoba
Winnipeg R3T 2N2
Canada .

Ruhr Universität
463 Bochum
W.Germany.

CENTRALIZERS IN LOCALLY FINITE GROUPS

B. Hartley
Department of Mathematics
University of Manchester
Manchester M13 9PL

By a locally finite group, we mean, of course, a group in which every finite set of elements generates a finite subgroup. In many problems on such groups, centralizers play a very important role. For example, they often provide a useful (and perhaps almost the only) way of constructing infinite proper subgroups and infinite abelian subgroups. More specifically, if G is an infinite simple locally finite group, then the Feit-Thompson Theorem tells us that G contains at least one involution i. The results to be described below ensure that $C_G(i)$ is infinite (Theorem 1.1) and cannot even be Černikov (Theorem 2.1). These can provide the main steps in proofs of the now well-known theorems that if G is locally finite and all its proper subgroups are finite then G is a quasicyclic group, (Hall-Kulatilaka [9], Kargapolov [19]) and if G is locally finite and satisfies the minimal condition on subgroups, then G is Černikov (Šunkov [30], Kegel-Wehrfritz [21]).

Thus, apart from their intrinsic interest, centralizers play a leading role in several areas of locally finite group theory.

In this paper, our theme will be the following. Given a locally finite group G, an element $x \in G$, and information about $C_G(x)$, what can we deduce about G? In some ways it is more appropriate to consider an automorphism α of G and investigate the effect on G of hypotheses on $C_G(\alpha)$. This allows the assumption that G has no non-trivial elements of order dividing the order of α to be introduced, an assumption which makes more tools available. We try to avoid this assumption as far as possible, however, as it is not applicable to the case when α is inner. We make no attempt to discuss the wealth of literature on fixed point free automorphisms of finite groups, but consider only two types of hypotheses:

(I.1) $C_G(\alpha)$ is finite

(I.2) $C_G(\alpha)$ is Černikov.

1. Finite centralizers

(1.1) Involutions. Partly because any two involutions in a group generate a dihedral subgroup, involutory automorphisms are usually the most tractable, so we begin with them. Most of the available information is contained in the following result, to which several authors have contributed. If $a,b,c,\ldots$ are numerical parameters, we use the term "$\{a,b,c,\ldots\}$-bounded" to mean "bounded above by some function of $a,b,c,\ldots$."

We have made no attempt to write down explicit bounds.

THEOREM 1.1. *Let* G *be a periodic group admitting an involutory automorphism* α *such that* $|C_G(\alpha)| = n < \infty$. *Then* G *contains a normal subgroup* H *such that* $|G : H|$ *is n-bounded and* $H' \leqslant C_G(i)$.

Since $|H'| \leqslant n$, we can consider $C_H(H')$ to deduce

COROLLARY. G *contains a normal subgroup of n-bounded index that is nilpotent of class at most two.*

The theorem appears as a direct generalization of the presumably well known fact that a periodic group P admitting a fixed point free involutory automorphism β, is abelian. For the proof of this, note that if $x \in P$, then the elements β and β^x of the semidirect product $P\langle\beta\rangle$ generate a dihedral group D of twice odd order. Hence $\beta^{xy} = \beta$ for some $y \in \langle\beta\beta^x\rangle$. We then find $xy = 1$, so $x = y^{-1}$ and $x^\beta = x^{-1}$.

An immediate consequence of the theorem is that G is locally finite. This part of the theorem, and in fact the most difficult part, is due to Šunkov (1972) [31], extending earlier work of his on periodic groups containing involutions, in which the centralizer of every involution is finite. The more general result is considerably deeper. We give a modified version of his proof in an appendix. He actually proves, by essentially brute-force methods, that G is almost soluble. As usual, we say a group is almost-$\underline{X}$, if it has an $\underline{X}$-subgroup of finite index.

However, once local finiteness has been established, a simple inverse limit argument along the lines of Kegel and Wehrfritz [20, p.54] reduces the proof of Theorem 1.1 to the finite case and it seems natural to use the methods of finite group theory. The argument then proceeds in four stages.

In (1.1A-D) below, G denotes a finite group with an automorphism α of order 2.

(1.1A) *If* $|C_G(\alpha)| \leqslant n$, *then* $|G : O_{2',2}(S(G))|$ *is n-bounded.*

Here $S(G)$ is the soluble radical of G, and the point of including it is to emphasise that this result does not depend on the Feit-Thompson Theorem. (1.1A) is due to Fong [7]. The proof is not very difficult. The case when G is simple is dealt with by a modification of the Brauer-Fowler argument, attributed to Goldschmidt, and the general case is deduced by considering the generalized Fitting subgroup.

(1.1B) *If* G *is soluble and* $|C_G(\alpha)| \leqslant n$, then $|G : F(G)|$ *is n-bounded.*

We denote the Fitting subgroup of a group X by $F(X)$.

The proof of (1.1B) involves analysing the action of the semidirect product $L = G\langle\alpha\rangle$ on its chief factors by a Hall-Higman type of approach involving a little representation theory. This is particularly simple because α is an involution, so the only representations that come into play are those of dihedral groups.

(1.1C) *If* G *is a finite q-group, where* q *is an odd prime, and* $|C_G(\alpha)| \leqslant q^m$, *then* G *contains an* α*-invariant subgroup* H *of nilpotency class at most two and index at most* $q^{f_1(m)}$, *where* $f_1(m) = 1^2 + 2^2 + \ldots + m^2$.

The proof of this is by induction on m, involving considering the action of α on the lower central factors of G and the use of commutator methods.

(1.1D) *If* G *is a finite 2-group and* $|C_G(\alpha)| \leqslant 2^m$, *then* G *contains a normal abelian subgroup of m-bounded index.*

We should perhaps remark that as long as we are not concerned with explicit bounds, it is irrelevant whether we look for subgroups of a particular type whose index is n-bounded, or α-invariant normal ones.

In proving (1.1D), it is actually shown that the index of an arbitrary maximal abelian normal subgroup A of $G\langle\alpha\rangle = L$ is m-bounded, by considering the action of α on A and the fact that $A = C_L(A)$.

The proofs of (1.1B-D) are in Hartley and Meixner [10]; see also [11] for a detailed account of all the above. It is a straightforward matter to assemble the above ingredients to prove the theorem.

A proof in the spirit of infinite group theory of the locally finite case of the corollary, but giving a normal nilpotent subgroup of class at most two and finite index rather than n-bounded index, has recently been given by Belyaev and Sesekin [5].

(1.2) Automorphisms of prime order

We consider now what happens to Theorem 1.1 and the various stages in its proof when α is allowed to be an automorphism of arbitrary prime order of G with a bounded number of fixed points. Firstly, there seems no hope of progress if G is merely assumed to be periodic rather than locally finite. This is because there is no analogue for odd primes of the fact that two involutions generate a dihedral group. Possibly something could be done by imposing some type of 2-generator condition on G. We also note that if p is an odd prime > 665, then there exist infinite p-groups with fixed point free automorphisms of order p. For if $B = \langle a,b\rangle$ is the free group on two generators in the Burnside variety defined by the law $x^p = 1$, then because of Adian's work [1], one knows that B is infinite and $C_B(a) = \langle a\rangle$. Now B/B' is elementary abelian of order p^2, so $\langle a\rangle \cap B' = 1$, and $C_{B'}(a) = 1$. Thus B' is the required group.

Thus we wish to consider the structure of a locally finite group admitting an automorphism of prime order p with a bounded number of fixed points, and for the type of results we have in mind, inverse limit arguments reduce the proof to the finite case. In (1.2A-D) below, G denotes a finite group with an automorphism α of prime order p.

To establish the analogue of (1.1A), we now require not only the Feit-Thompson Theorem, but also the full classification of finite simple groups, since the case when G is simple is dealt with by inspection. It may be worth remarking that the sporadic groups can be ignored in results of this type.

(1.2A) (CFSG). *If* $|C_G(\alpha)| \leqslant n$, *then* $|G:O_{p'p}(S(G))|$ *is* $\{p,n\}$*-bounded.*

We indicate by CFSG the fact that the classification of finite simple groups is involved in the proof, for which see Fong [7].

Next we have

(1.2B) *If* G *is finite soluble and* $|C_G(\alpha)| \leqslant n$ *then* $|G:F(G)|$ *is* $\{p,n\}$*-bounded.*

The proof is along the same lines as (1.1B), but the representation theory is a little more complicated, involving the representations of cyclic extensions of extra-special groups in the usual way. See Hartley and Meixner [12]; another proof under the assumption $p \nmid |G|$ has been given by Pettet.

It seems to be unknown whether the analogue of (1.1C) is true for automorphisms of odd prime order, or indeed what the correct analogue should be. The following special case has been obtained using Lie ring methods by Meixner [22], see also [23].

(1.2C) *Let* G *be a metabelian q-group, where* $q \neq p$, *and* $|C_G(\alpha)| \leq q^m$. *Then* G *contains a normal subgroup* H *of nilpotency class* $\leq 2p+1$ *and* $\{p,m,q\}$*-bounded index.*

It seems plausible that some analogue of this holds when G is not metabelian, with $2p+1$ replaced by some function of p (see [16] pp. 80-83 for some discussion of this). It is very straightforward to deduce from Higman's Theorem on fixed point free automorphisms of prime order [17] that if G is a finite q-group and $|C_G(\alpha)| \leq q^m$, then the derived length of G is $\{p,m\}$-bounded.

The analogue of (1.2D) has recently been established by Huhro [18] by a quite straightforward argument.

(1.2D) *Let* G *be a finite p-group and* $|C_G(\alpha)| \leq p^m$. *Then* G *has a subgroup of* $\{p,m\}$*-bounded index and nilpotency class* $\leq h(p)+1$, *where* $h(p)$ *is the "Higman function" bounding the nilpotency class of a finite group admitting a fixed point free automorphism of order* p.

There are examples showing that the bound $h(p)+1$ cannot be replaced by a constant independent of p and m (see [18], [16]); also Huhro gives an analogous result in which the index and the class are related to the derived length of G.

Huhro's argument runs as follows. It was observed by Alperin [2] that the proof of Higman's Theorem on fixed point free automorphisms yields the following:

If L *is a Lie ring admitting an automorphism of prime order* p *whose fixed points lie in an ideal* I *of* L, *then* $(pL)^{h(p)+1} \leq I$.

The exponent denotes the appropriate Lie power, and $h(p)$ is the "Higman function". We apply this to the Lie ring of G; since α has at most p^m fixed points on any α-invariant section of G, we can take $I = \{x \in L : p^m x\} = 0$. Thus we find that $p^{h(p)+m+1}$ annihilates $L^{h(p)+1}$. The lower central factors of G beyond the $h(p)+1$-st thus have bounded exponent. Their ranks are easily seen to be bounded, hence so are their orders. Let p^s be the bound obtained $(s = s(m,p))$ and $H = \gamma_{s+1}(G_1)$, where $G_1 = G\langle\alpha\rangle$. Now the lower central factors of G_1 have order at most p^m, since α operates trivially on them, so $|G : H|$ divides p^{ms}. Now a result of P. Hall ([8] Theorem 2.56) shows that all the non-trivial lower central factors of H, apart possibly from the last, have order at least p^{s+1}. Applying the above Lie ring considerations to H instead of G, we find that $\gamma_{h(p)+2}(H) = 1$, so H has class at most $h(p)+1$.

We can make a little more progress with the infinite version of (1.1C) for automorphisms of odd prime order by proving.

(1.2C') *Let* Q *be a locally finite q-group admitting an automorphism* α *of prime order* $p \neq q$ *such that* $C_Q(\alpha)$ *is finite. Then* Q *is hypercentral.*

Proof. It clearly suffices to prove that the centre $Z(Q)$ is non-trivial. If $C = C_Q(\alpha) = 1$, then Q is nilpotent by Higman's Theorem, so we may assume $C \neq 1$. Let N be a normal α-invariant subgroup of Q such that $C_1 = N \cap C$ is of minimal order subject to being non-trivial, and let M be the intersection of all α-invariant normal subgroups L of Q such that $L \cap C = C_1$. Thus $M \cap C = C_1$ and $M \neq 1$. If M is a minimal α-invariant normal subgroup of Q, then results essentially due to McLain (cf [20] p.12) show that $M \leq Z(Q)$. Thus we may assume that M contains a non-trivial proper α-invariant subgroup X. Then $C_X(\alpha) = 1$. Let Q_1 be the product of all normal α-invariant subgroups of Q on which α operates fixed point freely. We have shown that $Q_1 \neq 1$. Applying the above argument to Q/Q_1, we find that this group, if non-trivial, has a non-trivial centre which is also centralized by α. Using induction on $|C_Q(\alpha)|$, we find that Q has a finite series of α-invariant normal subgroups

$$1 < Q_1 < Q_2 < \dots < Q_r = Q.$$

such that $C_{Q_1}(\alpha) = 1$ and the remaining factors are either finite and central in $Q\langle\alpha\rangle$, or are transformed fixed point freely by α.

Now we prove by induction on i that $C_{Q_1}(Q_i) \neq 1$. If $i = 1$, this is because Q_1 is nilpotent. Assuming it holds for some value of i, the passage to $i+1$ is clear if Q_{i+1}/Q_i is finite. If not, let $C_i = C_{Q_1}(Q_i)$. Then α acts fixed point freely on the semidirect product $C_i \rtimes (Q_{i+1}/Q_i)$. This is then nilpotent by Higman's Theorem, so $C_{C_i}(Q_{i+1}) \neq 1$, as required.

Putting these various facts together gives the following omnibus theorem.

THEOREM 1.2. **(CFSG).** *Let* G *be a locally finite group admitting an automorphism* α *of prime order* p *with* $|C_G(\alpha)| \leq n$. *Then the Hirsch-Plotkin radical* $F(G)$ *of* G *has* $\{p,n\}$*-bounded index in* G, *and is hypercentral and soluble. Further,* $O_p(F(G))$ *has a subgroup of* $\{p,n\}$*-bounded index and nilpotency class* $\leq h(p)+1$, *and* $O_p(F(G))$ *is Černikov.*

The last assertion follows from Kegel and Wehrfritz ([20] 1.G.6), for example.

(1.3) Automorphisms of prime power order.

Now we consider a locally finite group G admitting an automorphism α of prime power order p^k with $|C_G(\alpha)| \leq n$. It does not seem clear at present whether the analogue of (1.2A) holds. However the "limiting" or "qualitative" version of it can be deduced from unpublished work of the author on the case when $C_G(\alpha)$ is Černikov.

(1.3A). [15] (CFSG). *Let* G *be a locally finite group admitting an automorphism* α *of prime power order such that* $C_G(\alpha)$ *is Černikov. Then* G *is almost locally soluble.*

Of course, this applies to the case when $|C_G(\alpha)| < \infty$. This result will receive further attention later.

Consider now the locally soluble case. For a group X, let $F(X)$ denote the Hirsch-Plotkin radical of X, and let $\{F_n(X)\}$ be the Hirsch-Plotkin series of X, defined by $F_0(X) = 1$ and $F_{i+1}(X)/F_i(X) = F(X/F_i(X))$. When X is finite, this is of course the usual Fitting series.

The following result is due to Meixner [22], [24].

(1.3B). *Let* G *be a finite soluble group admitting an automorphism* α *of prime power order* p^k *such that* $|C_G(\alpha)| \leq n$. *Then*

(i) *If* $p \neq 2$, $|G : F_k(G)|$ *is* $\{p,k,n\}$*-bounded*

(ii) *If* $p = 2$, *then* $|G: F_{2k-2}(G)|$ *is* $\{k,n\}$*-bounded.*

In fact, Meixner dealt with the case $p \nmid |G|$, but the extension to the general case has been carried out by the author (unpublished). It seems reasonable to expect that $2k-2$ can be replaced by k when $p = 2$. I know of no results on the structure of the Fitting factors $F_{i+1}(G)/F_i(G)$ along the lines of (1.2C), or (1.2D).

The results above give

THEOREM 1.3. *If* G *is a locally finite group admitting an automorphism* α *of prime power order* p^k, *with* $|C_G(\alpha)| < \infty$, *then* $|G : F_k(G)| < \infty$ *if* $p \neq 2$, *and* $|G : F_{2k-2}(G)| < \infty$ *if* $p = 2$.

Possibly G must be almost soluble, but I have no idea how to set about proving that.

2. Černikov centralizers

We recall that a group X is a Černikov group, if it has a normal subgroup X^0 of finite index that is a direct product of a finite number of groups of type C_{p^∞} for various primes p. Then X^0 is *divisible* in the sense that every element has an n-th root in it, for all $n \geq 0$. Note that X^0 can be characterized as the unique minimal subgroup of finite index of X, or the unique maximal divisible subgroup of X. By a deep result of Šunkov [30] (see also Kegel and Wehrfritz [21]), these are exactly the locally finite groups satisfying the minimal condition on subgroups, and much of the interest in elements with Černikov centralizers has its roots in Šunkov's work, as indeed do many of the methods. We note first the well known fact:

(2.0) *If* G *is locally finite and admits a finite p-group* P *of automorphisms such that* $C_G(P)$ *is Černikov, then* G *satisfies Min-p, the minimal condition on p-subgroups. Thus every p-subgroup of* G *is Černikov. (See Kegel and Wehrfritz* [20], 3.2).

It may be remarked that problems about Černikov centralizers do not reduce to the case of finite groups in any obvious way, unlike the problems about finite centralizers previously discussed.

1. Involutions

The most complete result here is due to Asar [4]. We state it in a form analogous to Theorem 1.1.

THEOREM 2.1. *Let* G *be a locally finite group admitting an involutory automorphism* α *such that* $C_G(\alpha)$ *is Černikov. Then* $G/[G,\alpha]$ *and* $[G,\alpha]'$ *are Černikov.*

Thus G departs from being abelian by a quotient and a normal subgroup whose structure reflect that of $C_G(\alpha)$. We easily deduce the following.

COROLLARY. G *is almost soluble*

Special cases of the above had been known previously. Šafiro and Šunkov [26] showed that a locally finite group G is almost locally soluble if $C_G(i)$ is Černikov for all involutions $i \in G$, and subsequently [27] extended this to the case when G contains a 4-group, each of whose involutions has Černikov centralizer. Pavlyuk has announced [25] a proof that with the hypothesis of Theorem 2.1, G is almost soluble.

The proof of 2.1 has several stages. By (2.0), G satisfies Min-2, so its Sylow 2-subgroups are Černikov (here the term "Sylow subgroup" is used in the sense of Wehrfritz [20]) so one can use induction on the "2-size" of G. The first stage is to show that $G/O_{2',2}(G)$ has finite Sylow 2-subgroups. This was done by Asar [3]. An error in his paper was pointed out and corrected by Stingl and Turau [29]. The locally soluble case of Theorem 2.1 is due to the author [13]. Its proof uses some simple representation theoretic ideas, which Asar has subsequently shown how to avoid. Unlike the proofs of the results on finite centralizers, the information on the locally soluble α-invariant subgroups plays a key rôle in Asar's proof of the general case, which is rendered more difficult by the fact that this information is not as strong as one would like. Roughly speaking, the proof consists of considering a counterexample G_1 in which the Sylow 2-subgroups have minimal order, and eventually extracting from G_1 a counterexample G which is essentially an infinite simple group. It should be noted that α could well be an outer automorphism. Further reductions lead to the discovery that PSL(2,K) is a counterexample, for a suitable infinite locally finite field K, and a contradiction ensues.

The classification of finite simple groups is not involved in this result, which indeed was proved before the classification was announced, but the traditional performers in this type of work, such as the Feit-Thompson Theorem, Bender's strongly embedded subgroup theorem, the Gorenstein-Walter Theorem, and one or two other results of that type, make their appearances, together with a theorem of Brauer on products of involutions, the proof of which depends on block theory.

The following variation on Theorem 2.1 may be worth stating (see [13]).

THEOREM 2.1'. *Let* G *be a locally finite group admitting an involutory automorphism* α *such that* $C_G(\alpha)$ *is Černikov. Then* G *contains a characteristic subgroup* K *such that*

(i) G/K *and* K' *are Černikov,*

(ii) K' *is abelian*

(iii) K' *is centralized by every involutory automorphism of* G *with Černikov centralizer.*

By analogy with the numbering of previous sections, we mention the following.

(2.1C) *Let* G *be a periodic locally nilpotent group admitting an automorphism* α *such that* $C_G(\alpha)$ *is Černikov. Then* $[G, \alpha]$ *is almost nilpotent.*

This follows easily from 2.1. For writing $H = [G, \alpha]$ we have that $K = H'$ contains a characteristic subgroup K^0 such that K/K^0 is finite and K^0 is an abelian Černikov group. Then $C_H(K^0) \cap C_H(K/K^0)$ is a nilpotent subgroup of finite index in H.

(2.2) Automorphisms of prime power order

We begin with the following unpublished result of the author, already stated as (1.3A).

(2.2A) [15] (CFSG) *Let* G *be a locally finite group admitting an automorphism* α *of prime power order such that* $C_G(\alpha)$ *is Černikov. Then* G *is almost locally soluble and so (in view of (2.0))* $G/O_{p'}(G)$ *is Černikov.*

The proof of this runs along the following lines. First, G has Min-p, by (2.0). The structure of groups with Min-p has been analysed by Wilson [35]. Combining his results with information following from CFSG, we find that G has a series

$1 \leqslant L \leqslant K \leqslant H \leqslant G$

of characteristic subgroups such that each of L and H/K is an extension of a p'-group by a p-group, G/H is finite, and K/L is a direct product of finitely many infinite simple groups of Lie type over locally finite fields of characteristic $\neq$ p. The classification of locally finite simple groups satisfying Min-p for some prime p, which involves studying direct limits of finite simple groups of Lie type and has been obtained independently by Borovik [6], Thomas [32], [33], and Hartley and Shute [14], is required here. This largely reduces the problem to showing that G cannot be an infinite simple group, which can be done on the basis of results of Steinberg on fixed points of automorphisms of algebraic groups [28] and the classification of algebraic groups defined over finite fields.

The case of (2.2A) when α has prime order has been obtained independently by Turau [34], using quite different methods.

As regards locally soluble groups, we have the following [13].

(2.2B) *Let* G *be a periodic locally soluble group admitting an automorphism* α *of prime order such that* $C_G(\alpha)$ *is Černikov, and let* F *be the Hirsch-Plotkin radical of* G. *Then*

(i) G/F *is Černikov*

and (ii) $[G, \alpha]F/F$ *is finite.*

We have no results about the case when α has prime power order. Nor have we been able to establish the analogue of (2.2C).

(2.3C) QUESTION. *Let* G *be a periodic locally finite q-group admitting an automorphism* α *of prime order* $p \neq q$ *such that* $C_G(\alpha)$ *is Černikov. Does it follow that* $[G, \alpha]$ *is hypercentral? (It can be shown that* G *has an ascending series of characteristic subgroups with abelian factors* [13]).

3. Appendix. Proof of Šunkov's Theorem

We state this explicitly as follows.

THEOREM 3.1. *Let* G *be a periodic group containing an involution* i *such that* $C_G(i)$ *is finite. Then* G *is locally finite.*

We have already noted that Šunkov proves that G is actually almost soluble. We shall not deal with this part of his work, preferring to regard it as belonging more naturally to the realm of finite group theory. We emphasize that the proof of Theorem 3.1 given below involves no ideas other than Šunkov's, but we feel it has some expositional advantages. We have deliberately kept our exposition as elementary and self contained as possible.

(3.2) *Let* G *be a locally finite p-group. If* G *contains an element* x *of order* p *with finite centralizer, then* G *is Černikov. If* $p = 2$, *then* x *inverts* G^0.

Proof. For the first statement, see [20]3.2. For the second, let $A = G^0$. If $a \in A$, then $a^2 = aa^{-i}.aa^{i}$, and i inverts aa^{-i} and centralizes aa^{i}. Thus as $A = A^2$, $A = [A,i]C_A(i)$. So $|A : [A,i]| < \infty$, and as A has no proper subgroup of finite index, $A = [A,i]$, which is inverted by i.

Remark. In fact, if G is a locally finite p-group, then either G is Černikov, or every finite subgroup of G centralizes an infinite elementary abelian subgroup of G.

(3.3) *Let* $G = A \rtimes \langle i \rangle$, *where* i *is an involution, and suppose* i *inverts* A (i.e. $iai = a^{-1}$ *for all* $a \in A$). *Then*

(i) A *is abelian,*

(ii) $G \setminus A$ *consists of involutions, which are all conjugate if* $A = A^2$.

We write $G = A \rtimes H$ to denote $G = AH$, $A \lhd G$, $A \cap H = 1$.

Proof. (i) If $a, b \in A$ then $(ab)^{-1} = b^{-1}a^{-1}$. Conjugating by i, we get $ab = ba$.

(ii) If $a \in A$ then $(ai)^2 = aiai = aa^{-1} = 1$. And $[a,i] = a^{-1}iai = a^{-2}$. If $b \in G \setminus A$ then as $A = A^2$, we have $b = a^{-2}i$ (for some $a \in A$) $= [a,i]i = i^a$.

(3.4) *Let* A *be a finite group of automorphisms of a group* G, *and let* N *be a normal locally finite A-invariant subgroup of* G. *Then*

$$C_{G/N}(A) = C_G(A)N/N$$

provided either

(a) N *has no non-trivial elements of order dividing* $|A|$, *or*

(b) $A = \langle i \rangle$, *where* i *is an involution inverting* $N = N^2$.

Proof. (a) This is a standard type of reduction to the finite case. Clearly $C_G(A)N/N \leq C_{G/N}(A) = C/N$ say. Conversely, let $H = C \rtimes A$. Then $NA \lhd H$. Let $x \in H$. Then $L = \langle A, A^x \rangle \leq NA$, so L is finite, and by the Dedekind law, $L = (L \cap N)A$. Now $(|L \cap N|, |A|) = 1$, so by the Schur-Zassenhaus Theorem (and the Feit Thompson Theorem) $A^{x\ell} = A$ for some $\ell \in L \cap N$. So $x\ell$ normalizes A, and $x \in N.N_H(A)$. Hence $H = N.N_H(A)$ and so $C = N.(C \cap N_H(A)) = N.C_C(A)$, as required.

(c) As before, $C_G(i)N/N \leq C/N = C_{G/N}(i)$. We have $N\langle i \rangle \lhd H = C \rtimes \langle i \rangle$, and if $h \in H$, then by (3.3) $i^{hn} = i$ for some $n \in N$. Thus $H = NC_H(i)$.

(3.5) *Let* α *be an automorphism of finite order of a group* G, N *be a normal locally finite* α*-invariant subgroup of* G *and suppose* $|C_G(\alpha)| = n < \infty$. *Then*

(i) *if either* N *has no non-trivial elements of order dividing the order* $|\alpha|$

of α, *or* α *is an involution inverting* $N = N^2$, *or* G *is locally finite, then* $|C_{G/N}(\alpha)| \leq n$.

(ii) *if* N *is finite, then* $|C_{G/N}(\alpha)| < \infty$.

Proof. (i) The first two cases follow from (3.4). In the third, if $|C_{G/N}(\alpha)| > n$, let F_1/N be a finite subgroup of $C_{G/N}(\alpha)$ with $|F_1/N| > n$. Let X be a transversal to N in F_1, and $F = \langle x^{\alpha^i} : x \in X, 0 \leq i < |\alpha|\rangle$. Thus F is finite and α-invariant and $F_1 = FN$. Let $M = F \cap N$. Then $[F,\alpha] \leq M$, and $|F/M| = |F_1/N| > n$. The map $\phi : x \longrightarrow [x,\alpha]$ maps F into M, and we see that $\phi(x) = \phi(y) \Leftrightarrow xy^{-1} \in C_F(\alpha)$. Thus $|\mathrm{im}\phi| = |F : C_F(\alpha)|$, so $|M| \geq |\mathrm{im}\phi|$ $= |F|/|C_F(\alpha)| \geq |F|/n$. Hence $|F/M| \leq n$, a contradiction.

(ii) $N\langle\alpha\rangle = L$ is a finite normal subgroup of $C\langle\alpha\rangle$ (where $C/N = C_{G/N}(\alpha)$) so $|C : C_C(N\langle\alpha\rangle)| < \infty$, and hence $|C : C_C(\alpha)| < \infty$.

(3.6) Question. To what extent can (3.5) be generalized to an arbitrary finite group of automorphisms?

Dihedral and locally dihedral groups

(3.7) *Let* $G = \langle i,j\rangle$ *where* $i^2 = j^2 = 1$ *and* $i,j \neq 1$, *and let* $a = ij$. *Then*

(i) $G = \langle a\rangle\langle i\rangle = \langle a\rangle\langle j\rangle$ *and* $b^t = b^{-1}$ *for all* $b \in \langle a\rangle$ *and* $t \in G\setminus\langle a\rangle$.

(ii) G *is finite if and only if* a *has finite order.*

(iii) *If* a *has odd order and* $t_1, t_2 \in G\setminus\langle a\rangle$ *then* $t_1^x = t_2$ *for some* $x \in \langle a\rangle$.

(iv) *If* a *has even order then* $\langle a\rangle$ *contains an involution which is central in* G *and is in particular centralized by* i *and* j.

These facts are of course all obvious. The significance is that because of (ii), we can obtain dihedral subgroups, that is, useful finite subgroups, in any periodic group containing more than one involution.

Let $C = \langle x_1, x_2, \ldots ; x_1^2 = 1, x_{i+1}^2 = x_i \ (i > 1)\rangle$ be a quasicyclic 2-group (Prüfer 2-group, group of type C_{2^∞}). C has an automorphism $i: c \rightarrow c^{-1}$ $(c \in C)$. The semi-direct product $D = C\langle i\rangle$ is the (unique up to isomorphism) *locally dihedral 2-group.*

(3.8) *With* D *as above, we have*

(i) *Every finite subset of* D *generates either a cyclic 2-group or a dihedral 2-group.*

(ii) *Every infinite subset of* C *generates* C *(because every proper subgroup of* C *is finite).*

(iii) *Every infinite subset of* $D\setminus C$ *generates* D.

Now we start the work of proving Theorem 3.1.

(3.9) *Let* i *be an involution in a periodic group* G *and suppose* $|C_G(i)| < \infty$. *Let* $\underline{\underline{M}}$ *be any infinite set of involutions in* G. *If* $|\langle is\rangle|$ *is even for infinitely many* $s \in \underline{\underline{M}}$, *then there exists an involution* $j \in C_G(i)$ *such that* $C_G(j) \cap \underline{\underline{M}}$ *is infinite.*

Proof. We may assume $|\langle is\rangle|$ is even for all $s \in \underline{\underline{M}}$. By 3.7 (iv), if $s \in \underline{\underline{M}}$, then $\langle i,s\rangle$ contains a central involution j_s. Thus each j_s belongs to $C_G(i)$, which is finite, and so there must be an infinite subset $\underline{\underline{M}}_0$ of $\underline{\underline{M}}$ such that all the

j_s ($s \in \underline{\underline{M}}_0$) coincide. This gives j.

(3.10) *Let* G *be a periodic group having an automorphism* α *of order* 2. *Suppose* G *contains a normal* α*-invariant subgroup* N *such that* α *inverts* N *and* G/N *and the Sylow 2-subgroup of* G/N *is divisible. Then* α *inverts* G *and* G *is abelian.*

Proof. We have $N \triangleleft G\langle\alpha\rangle$, N is abelian, and $G = N[G,\alpha]$, since every element of G/N has a square root. Now α is central in Aut N, so $[G,\alpha] \leq C_G(N)$. In other words, $N \leq Z(G)$, and G is nilpotent of class at most two. If $x,y \in G$ then $[x^\alpha, y^\alpha] = [x^{-1}, y^{-1}] = [x,y]$ as $x \equiv x^{-1}$, $y \equiv y^{-1} \bmod Z(G)$. But $[x^\alpha, y^\alpha] = [x,y]^\alpha = [x,y] \in N$. So $[x,y]^2 = 1$. But $x \equiv w^2 \bmod N$ for some $w \in G$, so $[x,y] = [w^2,y] = [w,y]^2 = 1$. Hence G is abelian. Also α inverts N and $[G,\alpha]$ and hence also G.

(3.11) *Let* G *be an infinite periodic group containing an involution* i *with finite centralizer. Then*

(i) i *inverts infinitely many elements of* G.

(ii) *Let* $g \in G$ *and let* $\underline{\underline{A}}$ *be any infinite set of elements of* G *inverted by* i. *Then either*

(iia) *Infinitely many elements* $i.i^{ag}$ ($a \in \underline{\underline{A}}$) *have even order. Then* $C_G(i)$ *contains an involution* j *such that* $C_G(j) \cap \{i^{ag} : a \in \underline{\underline{A}}\}$ *is infinite, or*

(iib) *Infinitely many elements* $i.i^{ag}$ ($a \in \underline{\underline{A}}$) *have odd order. Then there exists an element* $h \in C_G(i)$ *and an infinite subset* $\underline{\underline{A}}_0$ *of* $\underline{\underline{A}}$ *such that if* $a \in \underline{\underline{A}}_0$, *then* ia *is an involution inverting* gh^{-1}.

This technical lemma is one of the key results.

Proof. (i) Since G is infinite and $C_G(i)$ finite, i has infinitely many conjugates in G. Hence G contains infinitely many involutions j, and for each of these, i inverts ij.

(ii) In case (a), apply (3.9). So we may assume that $i.i^{ag}$ has odd order for all $a \in \underline{\underline{A}}$. Thus from (3.7), $i^{ags_a} = i$ for some $s_a \in \langle i.i^{ag}\rangle$; note that i inverts s_a. Hence $ags_a \in C_G(i)$, and we can choose $h \in C_G(i)$ and infinite subset $\underline{\underline{A}}_0$ of $\underline{\underline{A}}$ such that

$$ags_a = h \text{ for all } a \in \underline{\underline{A}}_0.$$

Conjugating by i gives $a^{-1}g^i s_a^{-1} = h$, and eliminating s_a gives

$$s_a = g^{-1}a^{-1}h = h^{-1}a^{-1}g^i.$$

Hence

$$h^{-1}a^{-1}g^i = g^{-1}iaih = g^{-1}iahi$$

and

$$h^{-1}a^{-1}ig = g^{-1}iah.$$

Rearranging, and noting that $ia = a^{-1}i$, and so ia is an involution, we have

$$(ia)(gh^{-1})(ia) = hg^{-1}$$

as required.

(3.12) *Let* i *be an involution with finite centralizer in an infinite periodic group* G. *Then* i *inverts an infinite abelian subgroup of* G.

Proof. First we claim that i inverts a non-trivial element y of G with infinite centralizer. For by (3.11), i inverts an infinite subset $\underline{\underline{A}}$ of G. Let $g \in G \setminus C_G(i)$, and apply 3.11 (ii). In case (iia) we have the desired conclusion. In case (iib), we obtain an element $h \in C_G(i)$ and an infinite subset $\underline{\underline{A}}_0$ of $\underline{\underline{A}}$ such that each $ia (a \in \underline{\underline{A}}_0)$ is an involution inverting gh^{-1}. Since $g \notin C_G(i)$, $gh^{-1} \neq 1$. If $a_1, a_2 \in \underline{\underline{A}}_0$, then $(ia_1)(ia_2) = a_1^{-1}a_2$ centralizes gh^{-1}, so $C_G(gh^{-1})$ is infinite. We do not know that i inverts gh^{-1}; if however some element i.ia $(a \in \underline{\underline{A}}_0)$ has odd order, then i is conjugate to ia and we have the desired conclusion. Otherwise, the products i.ia $(a \in \underline{\underline{A}}_0)$ all have even order, and (3.9) gives an involution $j \in C_G(i)$ such that $C_G(j)$ is infinite. So in all cases, the desired element y can be found.

Now we consider $N_G(\langle y \rangle)/\langle y \rangle$ and iterate the process. Suppose we have $1 = D_0 \lhd D_1 \lhd \ldots \lhd D_n$, where these are finite subgroups normalized by i, and i inverts each D_{r+1}/D_r, $D_n \leq N = \bigcap_{r=1}^{n} N_G(D_r)$, and N is infinite. Then $C_{N/D_n}(iD_n)$ is finite (3.5), so the first paragraph allows us to go one stage further. Eventually we obtain an infinite chain.

$$D_1 \leq D_2 \leq \ldots$$

with union D say, such that $D_r \lhd D\langle i \rangle$ for all r and i inverts each D_{r+1}/D_r.

Since the automorphism induced by i on each D_{r+1}/D_r commutes with every automorphism of that group, $E = [D,i]$ centralizes all the factors of the above series, and so is hypercentral. Since D is locally finite and i centralizes D/E, we have $|D : E| < \infty$ by (3.5), and E is infinite. Write

$$E = O_{2'}(E) \times O_2(E).$$

If $O_2(E)$ is infinite, then by (3.2), i inverts the infinite abelian group $O_2(E)^0$. Otherwise $O_{2'}(E)$ is infinite, and since it has an ascending series of finite normal i-invariant subgroups with factors inverted by i, repeated application of (3.10) shows that it is abelian and inverted by i.

Remark. Before proving Theorem 3.1, Šunkov had treated the case when all involutions of G have finite centralizer. In that situation, Case (iia) of (3.11) does not arise, the abelian group just obtained is a 2'-group, and quite a bit of the subsequent argument simplifies. This case is treated in [20].

(3.13) *Let* i *be an involution in an infinite periodic group* G, *such that* $|C_G(i)| < \infty$. *Let* A *be an infinite abelian subgroup of* G *inverted by* i, *and let* j *be an involution of* G. *If infinitely many elements* $j.i^a$ $(a \in A)$ *have odd order, then* j *inverts an infinite subgroup of* A.

Remark. The long term aim is to show that j inverts a subgroup of finite index of A, and then continue to more involutions.

Proof. Since certainly at least one $j.i^a$ has odd order, j is conjugate to i and so has finite centralizer. Write $j = i^{g^{-1}}$ $(g \in G)$. Then $i.i^{ag}$ has odd order for infinitely many $a \in A$. By (3.11), we can choose an infinite subset $\underline{\underline{A}}_0$ of A and an element $h \in C_G(i)$ such that if $a \in \underline{\underline{A}}_0$, then ia is an involution inverting gh^{-1}. Thus if $a_1, a_2 \in \underline{\underline{A}}_0$, then $a_1^{-1}a_2 \in C_A(gh^{-1})$, so this centralizer is infinite. Let $B = C_A(gh^{-1})$. Then i inverts B, so $j = i^{hg^{-1}}$ inverts $B^{hg^{-1}} = B$.

The next step gets control of the C_{2^∞}-subgroups inverted by i by showing that they all are normal in $\langle i^G \rangle$. The proof is quite striking.

(3.14) *Let* i *be an involution with finite centralizer in a periodic group* G *and suppose* C *is a* C_{2^∞}*-subgroup of* G *inverted by* i. *Let* $g \in G$. *Then* $gig^{-1} = i^{g^{-1}}$ *inverts* C.

<u>Proof</u>. Let $D = C\langle i \rangle$, a locally dihedral 2-group, and let

$$M = \langle i, D^g \rangle = \langle i, C^g, i^g \rangle$$
$$\langle i, C_1, i_1 \rangle$$

say, where $C_1 = C^g$ and $i_1 = i^g$. Recall that any infinite set of involutions i^a $(a \in C)$ generates D (3.8), so every infinite set of involutions $i_1^{g^{-1}ag} = i^{ag}$ $(a \in C)$, generates $C_1\langle i_1 \rangle$ and hence, together with i, generates M.

Let L be the largest normal locally finite 2-subgroup of M. By (3.2), L is Černikov, and i inverts L^0. Hence if $C_1 \leq L^0$, then i inverts C_1 and $i^{g^{-1}}$ inverts C. Suppose that $C_1 \not\leq L$, and let $\overline{M} = M/L$. Then $\overline{C}_1 = C_1L/L$ is infinite. Now $C_{\overline{M}}(\overline{i})$ is finite, as $C_{M/L^0}(i)$ is finite by 3.5(i), and then $C_{M/L}(i)$ is finite by 3.5(ii). We have an infinite set of involutions $\overline{i}_1^{\overline{a}}$ $(a \in C_1)$. If infinitely many elements $\overline{i}.\overline{i}_1^{\overline{a}}$ have even order, then by (3.9), $\overline{M}$ contains an involution centralized by $\overline{i}$ and infinitely many of the $\overline{i}_1^{\overline{a}}$. These elements generate $\overline{M}$, so $\overline{M}$ contains a central involution, contrary to the choice of L. If however infinitely many of the $\overline{i}.\overline{i}_1^{\overline{a}}$. These elements generate $\overline{M}$, so $\overline{M}$ contains a central involution, contrary to the choice of L. If however infinitely many of the $\overline{i}.\overline{i}_1^{\overline{a}}$ have odd order, then by (3.13), $\overline{i}$ inverts $\overline{C}_1$. This gives $\overline{C}_1 \triangleleft \overline{M}$, again a contradiction to the choice of L. This concludes the proof of (3.14).

We could now factor out all the groups like C, but defer that for the moment.

(3.15) *Let* i,k *be involutions with finite centralizer in an infinite periodic group* G. *Suppose that* i *inverts an infinite abelian* $2'$*-subgroup* A *of* G, *and that* k *normalizes no* C_{2^∞}*-subgroup of* G. *Then* k *inverts a subgroup of finite index of* A.

<u>Proof</u>. <u>Step 1. k inverts an infinite subgroup of A</u>. If k were to normalize an infinite locally finite 2-subgroup R of G, then by (3.2), R would be Černikov and k would invert R^0, contrary to hypothesis. So we can choose a finite 2-subgroup T of G, maximal subject to being normalized by k and an infinite subset $\underline{\underline{N}}$ of $\underline{\underline{M}} = \{i^a : a \in A\}$. Let $i_1 \in \underline{\underline{N}}$. Then $\langle i_1 \rangle \leq \langle \underline{\underline{N}} \rangle \leq A\langle i_1 \rangle$, so $\langle \underline{\underline{N}} \rangle = A_1\langle i_1 \rangle$ say, where $A_1 = A \cap \langle \underline{\underline{N}} \rangle$, and the members of $\underline{\underline{N}}$ are conjugate to i_1 under A_1. Let $N = N_G(T)$, which contains $\langle k, A_1, i_1 \rangle$, and $\overline{N} = N/T$. The set $\underline{\underline{\overline{M}}} = \langle \overline{i}_1^{\overline{a}} : \overline{a} \in \overline{A}_1 \rangle$ is infinite (use (3.5)). Consider the products $\overline{k}.\overline{i}_1^{\overline{a}}$ $(\overline{a} \in \overline{A}_1)$. If infinitely many have even order, then (3.9) gives us an involution in $\overline{N}$ centralized by $\overline{k}$ and infinitely many $\overline{i}_1^{\overline{a}}$ $(\overline{a} \in \overline{A}_1)$, and we obtain a contradiction to the maximality of T. So infinitely

many have odd order, and (3.13) gives us an infinite subgroup $\overline{B} = B/T$ of A_1T/T, inverted by $\overline{k}$. Let $C = C_B(T)$, which has finite index in B and so is infinite. Then C is nilpotent of class at most two, so $C = O_2(C) \times O_{2'}(C)$, since A_1 is a maximal 2'-subgroup of A_1T, we have $O_{2'}(C) \leq A_1$, and clearly k inverts $O_{2'}(C)$.

Step 2. General case. Let D be the largest subgroup of A inverted by k, and $N = \langle A,i,k\rangle \leq N_G(D)$. Since D is a 2'-group, if $\overline{N} = N/D$ we have $|C_{\overline{N}}(\overline{k})| < \infty$ (3.4); also if $\overline{C} = C/D$ is a C_{2^∞}-subgroup of $\overline{N}$ inverted by $\overline{k}$, then by (3.10), k inverts C, which is abelian, and k inverts the C_{2^∞}-subgroup $O_2(C)$ of G, contrary to supposition. So if $\overline{A}$ is infinite, Step 1 gives us an infinite subgroup $\overline{B} = B/D$ of $\overline{A}$ inverted by k, and then k inverts B, contradicting the choice of D. Hence $|A:D| < \infty$.

Finally we come to

Proof of Theorem 3.1 Let $H = \langle i^G\rangle$. Every C_{2^∞}-subgroup of G normalized by i is inverted by i and contained in H, and so, by (3.14), it is normalized by every conjugate of i and is normal in H. Thus any two such subgroups normalize each other and hence commute elementwise. The join R of all such subgroups is the unique maximal normal divisible 2-subgroup of H. If $\overline{G} = G/R$, then $C_{\overline{G}}(\overline{i})$ is finite (3.5) and $\overline{i}$ inverts no C_{2^∞}-subgroup of $\overline{G}$. Passing to $\overline{G}$, we may assume without loss of generality that $R = 1$.

We shall show first that H is locally finite. Evidently we may suppose that G is infinite, so (3.12) gives us an infinite abelian subgroup A inverted by i. Since i normalizes no C_{2^∞}-subgroup of G, $O_2(A)$ is finite, so we may take A to be a 2'-group, and then choose A to be maximal among the infinite abelian 2'-subgroups of G inverted by i.

Let $g_1,\ldots,g_n \in G$. Applying (3.15) several times, we obtain a subgroup B, of finite index in A, inverted by all the involutions i, $i^{g_1},\ldots, i^{g_n}$. All these involutions are congruent mod $C_G(B)$, and so lie in $L = C_G(B)\langle i\rangle$.

We claim that $\overline{L} = L/B$ is finite. Since B is abelian, this will imply that L is locally finite, so $\langle i^{g_1},\ldots, i^{g_n}\rangle$ will be finite and H locally finite. So assume that $\overline{L}$ is infinite. Then $C_{\overline{L}}(\overline{i})$ is finite, so $\overline{i}$ inverts an infinite abelian subgroup $\overline{D} = D/B$ of $\overline{L}$. Using (3.10), we see that $\overline{D}$ can be taken to be a 2'-group, and i inverts D. We almost have a contradiction to the maximality of A. To obtain one, note that as $|A: B| < \infty$, we can choose a finite subgroup F of A, necessarily inverted by i, such that $A = BF$. Now $\langle F,i\rangle$ is generated by finitely many conjugates of i, each of which inverts a subgroup of finite index of D, by (3.15). So F centralizes a subgroup D_0 of finite index of D, $A = BF$ centralizes D_0 and AD_0 is abelian and inverted by i. We *have* a contradiction to the maximality of A.

We now know that H is locally finite. Let S be a maximal 2-subgroup of H containing i. Then S is Černikov and i inverts S^0. So $S^0 = 1$ and S is finite. Therefore the Sylow 2-subgroups of H are conjugate in H, and we can use the Frattini argument to give $G = HN_G(S)$. Clearly $|N_G(S) : C_{N_G(S)}(i)|<\infty$, so $N_G(S)$

is finite. Finally $|G : H| < \infty$, so G is locally finite

References

1. S. Adian, The Burnside problem and identities in groups, trans. John Lennox and James Wiegold, Springer-Verlag, Berlin, New York, 1979.
2. J.L. Alperin, "Automorphisms of solvable groups", Proc. Amer. Math. Soc. 13 (1962), 175-180.
3. A.O. Asar, "On a problem of Kegel and Wehrfritz", J. Algebra 59 (1979), 47-55.
4. A.O. Asar, "The solution of a problem of Kegel and Wehrfritz", Proc. London Math. Soc. (3) 45 (1982), 337-364.
5. V.V. Belyaev and N.F. Sesekin, "Periodic groups with an almost regular involutive automorphism," Studia Sci. Math. Hungar. 17 (1982), 137-141.
6. A.V. Borovik, "Embeddings of finite Chevalley groups and periodic linear groups," Sibirsk. Mat. Zh. 24 (1983), 26-35 (Russian) = Siberian Math. J. 24 (1983), 843-55.
7. P. Fong, "On orders of finite groups and centralizers of p-elements," Osaka J. Math. 13 (1976), 483-489.
8. P. Hall, "A contribution to the theory of groups of prime power order," Proc. London Math. Soc. (2) 36 (1934), 29-95.
9. P. Hall and C.R. Kulatilaka, "A property of locally finite groups," J. London Math. Soc. 39 (1964), 235-239.
10. B. Hartley and Th. Meixner, "Periodic groups in which the centralizer of an involution has bounded order," J. Algebra 64 (1980), 285-291.
11. B. Hartley, Centralizers in locally finite groups, Lecture Notes No. 12, Department of Mathematics, National University of Singapore, 1982.
12. B. Hartley and Th. Meixner, "Finite soluble groups in which the centralizer of an element of prime order is small," Arch. Math. 36 (1981), 211-213.
13. B. Hartley, "Periodic locally soluble groups containing an element of prime order with Černikov centralizer," Quart. J. Math. Oxford (2) 33 (1982), 309-323.
14. B. Hartley and G. Shute, "Monomorphisms and direct limits of finite groups of Lie type," Quart. J. Math. Oxford (2) 35 (1984), 49-71.
15. B. Hartley, "Fixed points of automorphisms of prime power order of locally finite groups and Chevalley groups", in preparation.
16. B. Hartley, Topics in the theory of nilpotent groups, in Group Theory. Essays for Philip Hall (Academic Press, London, 1984).
17. G. Higman, "Groups and rings having automorphisms without non-trivial fixed points," J. London Math. Soc. 32 (1957), 321-334.
18. E.I. Huhro, "Finite p-groups admitting an automorphism of order p with a small number of fixed points," Mat. Zametki 38 (1985), 652-657 (Russian).
19. M.I. Kargapolov, "On the problem of O. Yu. Šmidt, Sibirsk. Mat. Zh. 4 (1963), 233-235.
20. O.H. Kegel and B.A.F. Wehrfritz, Locally finite groups (North Holland, Amsterdam, 1973).
21. O.H. Kegel and B.A.F. Wehrfritz, "Strong finiteness conditions in locally finite groups," Math. Z. 117 (1970), 309-324.
22. Th. Meixner, Über endliche Gruppen mit Automorphismen, deren Fixpunktgruppe beschränkt sind (Doctoral thesis, Universität Erlangen-Nürnberg, 1979).
23. Th. Meixner, "Metabelsche Gruppen mit einem fixpunktfreien Automorphismus von Primzahlordnung," Arch. Math. 35 (1980), 497-500.
24. Th. Meixner, "Solvable groups admitting an automorphism of prime power order whose centralizer is small," J. Algebra 99 (1986), 181-190.
25. I.I. Pavlyuk, "On a problem of Kegel and Wehrfritz", Proc. Seventh All-Union Symposium on Group Theory, Krasnoyarsk, 1980.
26. A.A. Šafiro and V.P. Šunkov, "Locally finite groups with Černikov centralizers of involutions," in Studies in Group Theory, Inst. Fiz. im Kirenskogo Sibirsk. Otdel Akad. Nauk SSSR, Krasnoyarsk, 1975.
27. A.A. Šafiro and V.P. Šunkov, "A characterization of an infinite Černikov group which is not a finite extension of a quasicyclic group," Mat. Sb. (N.S.) 107 (149)(1978), 289-303.
28. R. Steinberg, Endomorphisms of algebraic groups, Mem. Amer. Math. Soc. 80 (1968).
29. V. Stingl and V. Turau, "Locally finite Syl_p^*-groups," J. Algebra 90 (1984), 354-363.

30. V.P. Šunkov, "On the minimality problem for locally finite groups," Algebra i Logika 9 (1970), 220-248 (Russian) = Algebra and Logic 9 (1970), 137-151.
31. V.P. Šunkov, "On periodic groups with an almost regular involution," Algebra i Logika 11 (1972), 470-493 (Russian) = Algebra and Logic 11 (1972), 260-272.
32. Simon Thomas, "An identification theorem for the locally finite nontwisted Chevalley groups," Arch. Math. 40 (1983), 21-31.
33. Simon Thomas, "The classification of the simple periodic linear groups," Arch. Math. 41 (1983), 103-116.
34. V. Turau, "Zentralisatoren in lokal-endlichen Gruppen vom Chevalley-Typ," Arch. Math. 44 (1985), 297-308.
35. J.S. Wilson, "On groups satisfying Min-p," Proc. London Math. Soc. (3) 26 (1973), 226-248.

SUBGROUP EMBEDDING PROPERTIES

by Trevor Hawkes

University of Warwick
Coventry CV4 7AL, England

In group theory, many of the theorems are statements about the existence and properties of certain kinds of subgroups - the theorems of Lagrange and Sylow, and the p^aq^b-theorem of Burnside are classical examples; almost all of the proofs involve some analysis of subgroup structure. What follows is a preliminary approach to the problem of studying systematically the different ways a subgroup can be embedded in a group. Just as the calculus of closure operations provides a helpful viewpoint for dealing with classes of groups, so I feel there is scope for developing a formal framework in which to present, compare, and contrast the various subgroup embedding properties met with in group theory.

From the beginning, normal subgroups were of central concern because of their role in classifying group-homomorphisms. Subnormal subgroups have also been closely studied; these are the terms of composition series, whose significance is clear from the Jordan-Hölder theorem. Investigations of groups with a rich subgroup structure, such as soluble groups, have produced a diverse assortment of subgroup embedding properties, many of which generalize properties of Sylow and Hall subgroups. Here are a few examples, each reflecting some aspect of the behaviour of Sylow subgroups. Let H be a subgroup of a finite group G. Then we say

H is <u>pronormal</u> in $G \Leftrightarrow H$ and H^g are conjugate in $\langle H, H^g \rangle$ for all $g \in G$,

H <u>satisfies the Frattini argument</u> in $G \Leftrightarrow G = KN_G(H \cap K)$ for all $K \trianglelefteq G$,

H is a cover-avoidance property (CAP) subgroup $\Leftrightarrow$ H either covers or avoids abelian chief factors, and finally

H is system permutable in a soluble group G $\Leftrightarrow$ G has a Hall system Σ such that $HS = SH$ for all $S \in \Sigma$.

These examples may be used to test out the ideas introduced below.

Any general mathematical theory should be judged by whether it unifies, or at least illuminates, existing knowledge, making new connections and giving new insights; and by whether it raises new questions and suggests fruitful new directions. I hope that the tentative suggestions outlined in the sequel will provoke some interest in the question whether a general framework for studying embedding properties is either a feasible, or a desirable, undertaking.

§1. General Embedding Properties

DEFINITION 1.1. A subgroup embedding property (SEP) is a map $\mathcal{E}$ which associates with each group G (in some fixed universe) a subset $\mathcal{E}(G)$ of $\mathcal{S}(G)$, the set of all subgroups of G, and satisfies

$$(*) \qquad \alpha(\mathcal{E}(G)) = \mathcal{E}(\alpha(G))$$

for all group-isomorphisms $\alpha : G \to \alpha(G)$.

This definition is clearly very general, and represents merely the minimum requirement that an SEP should be an invariant of each isomorphism class of groups. Most useful embedding properties satisfy additional conditions, for example one or more of the following:

(a) Q-invariance: $\mathcal{E}$ is quotient-invariant if equation (*) holds for all epimorphisms $\alpha : G \to \alpha(G)$.

Subnormality and pronormality are both examples of Q-invariant SEP's. In fact, these two examples each satisfy the conditions:

(i) Whenever $N \trianglelefteq G$ and $H \in \mathcal{E}(G)$, then $NH \in \mathcal{E}(G)$;

(ii) Whenever $N \trianglelefteq G$ and $N \leq H \in \mathcal{E}(G)$, then $H/N \in \mathcal{E}(G/N)$.

Together, these conditions clearly ensure that $\mathcal{E}(G)$ is Q-invariant.

(b) S-<u>invariance</u>: $\mathcal{E}$ is <u>subgroup-invariant</u> if, for all $S \leq G$, $\mathcal{E}(S) = \{ S \cap H \mid H \in \mathcal{E}(G) \}$.

This is a demanding requirement and is not often met. It is fulfilled when $\mathcal{E}(G)$ is the set of p-subgroups G, but is not satisfied by such embedding properties as subnormality or pronormality. A less stringent and more useful concept is the following:

(c) <u>Persistence</u>: $\mathcal{E}$ is <u>persistent</u> if, for all G and all $S \leq G$, we have $H \in \mathcal{E}(S)$, whenever $H \leq S$ and $H \in \mathcal{E}(G)$.

(d) s_n-<u>invariance</u>: $\mathcal{E}$ is <u>subnormal-subgroup-invariant</u> if, for all S sn G, $\mathcal{E}(S) = \{ S \cap H \mid H \in \mathcal{E}(G) \}$.

Embedding properties which are s_n-invariant are called <u>Fitting functors</u> and have been extensively studied by Beidleman, Brewster and Hauck [2].

(e) <u>Lattice-closure</u>: $\mathcal{E}$ is <u>lattice-closed</u> if $\mathcal{E}(G)$ is a sublattice of $\mathcal{L}(G)$ for all groups G.

Normality and subnormality are examples of lattice-closed embedding properties. Pronormality is not lattice-closed, although it displays some lattice-like behaviour in the universe of finite soluble groups, as we shall see below.

We end this section on general embedding properties by describing some ways of combining or modifying known embedding properties to obtain new ones. Embedding properties admit an obvious partial order defined as follows:

$\mathcal{E} \leq \mathcal{E}^*$ if and only if $\mathcal{E}(G) \subseteq \mathcal{E}^*(G)$ for all groups G; moreover they form a lattice with respect to this partial order with join and meet operations defined thus:

(i) $(\mathcal{E} \vee \mathcal{E}^*)(G) = \mathcal{E}(G) \cup \mathcal{E}^*(G)$, and

(ii) $(\mathcal{E} \wedge \mathcal{E}^*)(G) = \mathcal{E}(G) \cup \mathcal{E}^*(G)$ for all groups G.

These definitions of evidently extend to the join and meet of arbitrary sets of embedding properties. Another kind of join is the embedding property $\langle E_\lambda \mid \lambda\in\Lambda\rangle$ <u>generated by</u> the embedding properties $\{\mathcal{E}_\lambda\}$, defined thus

(iii) $\langle E_\lambda \mid \lambda\in\Lambda\rangle(G) = \{ \langle H_\lambda \mid \lambda\in\Lambda\rangle \mid H_\lambda\in\mathcal{E}_\lambda(G) \}$.

The <u>composition</u> $\mathcal{E}\circ\mathcal{E}^*$ of two embedding properties $\mathcal{E}$ and $\mathcal{E}^*$ is defined by setting

(iv) $(\mathcal{E}\circ\mathcal{E}^*)(G) = \{ L\in\mathcal{E}(H) \mid H\in\mathcal{E}^*(G) \}$.

An embedding property, such as subnormality, which satisfies $\mathcal{E}\circ\mathcal{E} = \mathcal{E}$ is called <u>transitive</u>. Since the intersection of transitive embedding properties is clearly transitive, we can form the transitive closure $\mathcal{E}^T$ of an embedding property $\mathcal{E}$ by

(v) $\mathcal{E}^T = \wedge\{ \mathcal{D} \mid \mathcal{D}$ transitive and $\mathcal{D} \geq \mathcal{E} \}$.

Since maximal subgroups of a finite group are always pronormal, the transitive closure of pronormality is the universal embedding property $\mathcal{S}:G \rightarrow \mathcal{S}(G)$.

We can also define the lattice closure $\mathcal{E}^L$ of $\mathcal{E}$ as follows:

(vi) $\mathcal{E}^L(G) =$ the sublattice of $\mathcal{L}(G)$ generated by $\mathcal{E}(G)$.

If $\mathcal{E}(G) = \{$pronormal subgroups of $G\}$, when G is nilpotent we have $\mathcal{E}(G) = \{$normal subgroups of $G\} = \mathcal{E}^L(G)$, and so $\mathcal{E}^L \neq \mathcal{S}$ in this case.

§2. <u>Lattice-related embedding properties</u>

Most subgroup embedding properties encountered in group theory are compounded from the following three sources of information about the set $\mathcal{S}(G)$:

(A) Lattice Structure, i.e. information contained in the subgroup lattice $\mathcal{L}(G) = (\mathcal{S}(G), \wedge, \vee)$, viewed as an abstract lattice.

(B) Conjugacy, i.e. information contained in $\{ \mathcal{C}_H(G) \mid H \leq G \}$, where $\mathcal{C}_H(G)$ is the pair $(\mathcal{S}(G), \sim_H)$, and $\sim_H$ denotes the equivalence relation of H-conjugacy on the set $\mathcal{S}(G)$.

(C) Order, i.e. knowledge of $\mathcal{O}(G) = (\mathcal{S}(G), \omega)$, where ω is the map $G \to |G|$.

The property $\mathcal{E}(G)$ = {maximal subgroups of G} is visible from the lattice; normality is determined from knowledge of $\mathcal{C}_G(G)$; and the property of being a Hall subgroup depends only on $\mathcal{O}(G)$. More often than not, though, a combination of these three types of information is required to define an embedding property. For example, all three are required to identify a strongly embedded maximal subgroup (viz. one of even order which intersects its conjugates in subgroups of odd order).

Next we consider the problem of giving precise meaning to the statement "$\mathcal{E}(G)$ is discernable from the subgroup lattice $\mathcal{L}(G)$".

If $\mathcal{L} = \mathcal{L}(G)$, $\mathrm{Aut}(\mathcal{L})$ will denote the group of lattice automorphisms of $\mathcal{L}$. If $\alpha \in \mathrm{Aut}(G)$, clearly α induces an automorphism $\bar{\alpha}$ on $\mathcal{L}$, and the map $\alpha \to \bar{\alpha}$ is a homomorphism from Aut(G) into Aut(L) with kernel Pot(G), the group of power automorphisms of G. In general, $\overline{\mathrm{Aut}(G)}$ is a proper subgroup of Aut(L) and, of course, there may be more than one group G with $\mathcal{L}(G)$ isomorphic with a given lattice $\mathcal{L}$.

DEFINITION 2.1. A subgroup embedding property $\mathcal{E}$ is lattice-stable if, whenever G is a group for which there exists a lattice isomorphism $\phi : \mathcal{L}(G) \to \mathcal{L}$, then $\phi(\mathcal{E}(G))$ is invariant under the action of $\mathrm{Aut}(\mathcal{L})$ on $\mathcal{L}$.

Example. Let $\mathcal{L}$ = 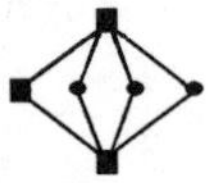 . Then $\mathrm{Aut}(\mathcal{L}) \cong S_4$, the symmetric group of degree 4. There exist two groups whose subgroup lattices are isomorphic with $\mathcal{L}$, namely $Z_3 \times Z_3$ and S_3. Since $\mathrm{Aut}(Z_3 \times Z_3) = GL(2,3)$ and $\mathrm{Pot}(Z_3 \times Z_3) = Z(GL(2,3))$, and since $PGL(2,3) \cong S_4$, the image of $\mathrm{Aut}(Z_3 \times Z_3)$ under the bar map is $\mathrm{Aut}(\mathcal{L})$. However, $\mathrm{Aut}(S_3) \cong S_3$ and $\mathrm{Pot}(S_3) = 1$, so in this case the bar map sends $\mathrm{Aut}(S_3)$ to one of the point stabilizers in $\mathrm{Aut}(\mathcal{L})$. The normal (and subnormal) subgroups of S_3, under an appropriate isomorphism from $\mathcal{L}(S_3)$ to $\mathcal{L}$, map to the square nodes in the diagram, and these are clearly not $\mathrm{Aut}(\mathcal{L})$-invariant. Thus neither normality nor subnormality are lattice-stable properties.

Example. The embedding property $\mathcal{E}(G) = \{$maximal subgroups of $G\}$ is obviously lattice-stable. This is a special case of a family of lattice-stable properties defined as follows: If $H \leq G$, let $\mathcal{L}(H,G)$ denote the sublattice of $\mathcal{L}(G)$ comprising subgroups which contain H. Let $\underline{\underline{L}}^{\uparrow}$ and $\underline{\underline{L}}^{\downarrow}$ be two classes of lattices. Then define

$$\mathcal{E}_{\underline{\underline{L}}}(G) = \{\ H \leq G \mid \mathcal{L}(H) \in \underline{\underline{L}}^{\downarrow} \text{ and } \mathcal{L}(H,G) \in \underline{\underline{L}}^{\uparrow}\ \} .$$

It is at once clear from the definition that for arbitrary pairs $(\underline{\underline{L}}^{\uparrow}, \underline{\underline{L}}^{\downarrow})$ the map $\mathcal{E}_{\underline{\underline{L}}}$ defines a subgroup embedding property. For example, by taking $\underline{\underline{L}}^{\uparrow}$ to the class of lattices with 2 elements and $\underline{\underline{L}}^{\downarrow}$ the class of all lattices, we find that $\mathcal{E}_{\underline{\underline{L}}}(G) = \{$maximal subgroups of $G\}$.

DEFINITION 2.2 We say a subgroup H of G is catenary if $H \in \mathcal{E}_{\underline{\underline{L}}}(G)$ when $\underline{\underline{L}}^{\uparrow} = \underline{\underline{L}}^{\downarrow}$ is the class of chains.

Examples. 1) If $W = Z_p \wr_{reg} Z_p$ and B denotes its base group, then $\langle x \rangle$ is a catenary subgroup whenever $x \in W \setminus B$.

2) The subgroup $\langle(1234)\rangle$ is catenary in S_4.

3) The subgroup $\langle(12345)\rangle$ is catenary in S_5.

PROPOSITION 2.3. Let G be a finite soluble group with a catenary subgroup. Then there exist distinct primes p and q such that $G \in \underline{\underline{S}}_p \underline{\underline{S}}_q \underline{\underline{S}}_p$; in particular, G has nilpotent length at most 3.

Conjecture. Let G be a finite group with a catenary subgroup. Then G has generalized nilpotent length at most 3.

A less stringent condition than lattice-stability is the following.

DEFINITION 2.4. A subgroup embedding property $\mathcal{E}$ is called lattice-invariant if, whenever G and H are groups with $\mathcal{L}(G) \cong \mathcal{L}(H)$, there exists an iosmorphism $\phi:\mathcal{L}(G) \to \mathcal{L}(H)$ such that $\phi(\mathcal{E}(G)) = \mathcal{E}(H)$.

Example It is clear from the earlier example, with $G = Z_3 \times Z_3$ and $H = S_3$, that normality and subnormality both also fail to be lattice-invariant. It is not difficult to devise an example, admittedly artificial, to show that the concept of lattice-invariance embraces more than lattice-stability. For let H be a group determined by its subgroup lattice, and let $\mathcal{X}$ be an Aut(H)-invariant subset of $\mathcal{L}(H)$ which is not invariant under $\mathrm{Aut}(\mathcal{L}(H))$. Set $\mathcal{E}(H) = \mathcal{X}$ and $\mathcal{E}(G) = \{1\}$ when $G \not\cong H$. Then $\mathcal{E}$ is lattice-invariant but not lattice-stable. [If H is the holomorph of Z_7, $\mathrm{Aut}(\mathcal{L}(H))$ contains an involution which interchanges the Sylow 2- and Sylow 3-subgroups. Therefore we could take $\mathrm{Syl}_2(H)$ for $\mathcal{X}$ in this example.]

§3. Conjugacy- and order-related properties

In the previous section we considered properties invariant under the lattice-preserving group of permutations of $\mathcal{S}(G)$ (the set of all subgroups of G). The "normal structure" of $\mathcal{S}(G)$ is perhaps best reflected by the directed graph $\Gamma = \Gamma(G)$ defined, for $K,L \in \mathcal{S}(G)$, by

$$(K,L) \in \Gamma \iff L \leq N_G(K) \ .$$

This gives rise to an associated subgroup $A^{\triangleleft}(G)$ of $\mathrm{Sym}(\mathcal{S}(G))$ comprising permutations which leave Γ invariant. The subgroup $A^{\sim}(G)$ preserving the relation of G-conjugacy on $\mathcal{S}(G)$ is also of interest here, although it is not clear how to define a subgroup that simultaneously preserves the more complicated set of structures $\mathcal{C}_H(G)$, when H runs through the subgroups of G. It is easy to find examples with $A^{\triangleleft}(G) \not\leq A^{\sim}(G)$ and vice-versa, and in any case these groups are probably too large to carry significant information about the subgroup-structure, but only become more interesting when their domains are restricted to $\mathrm{Aut}(\mathcal{L})$ (where $\mathcal{L} = \mathcal{L}(G)$), when they give rise to groups denoted by $\mathrm{Aut}^{\triangleleft}(\mathcal{L})$ and $\mathrm{Aut}^{\sim}(\mathcal{L})$. Embedding properties such as normality and subnormality are clearly invariant under both $\mathrm{Aut}^{\triangleleft}$ and $\mathrm{Aut}^{\sim}$. The group $\mathrm{Aut}^{\omega}(\mathcal{L})$ of lattice automorphisms which preserve the order map $\omega : G \to |G|$ (known as index-preserving autoprojectivities) has been extensively studied and leaves invariant embedding properties defined in terms of subgroup orders or indices, such as Hall subgroups. Of course, it is well known that there exist non-isomorphic groups G and H for which there exists a lattice isomorphism $\phi : \mathcal{L}(G) \to \mathcal{L}(H)$ preserving conjugacy, the normality graph Γ, and the order map ω. But in the study of subgroup embedding properties such groups are indistinguishable (i.e. one has $\phi(\mathcal{E}(G)) = \mathcal{E}(H)$ for all naturally occuring embedding properties $\mathcal{E}$), so they should be viewed as equivalent in this context.

To conclude this proposal for a framework for studying subgroup embedding properties, we move away from generalities and discuss some concrete examples that have been useful in the theory of soluble groups. We confine our attention to finite groups, and focus on a circle of properties which mimic aspects of the behaviour of Sylow subgroups. It is well known that for a finite group G,

(*) if $P \in \mathrm{Syl}_p(G)$ and K sn G, then $P \cap K \in \mathrm{Syl}_p(K)$.

What meaning can be given to the statement that an arbitrary subgroup H satisfies a condition like (*)? The key to answering this question is to find a set $\mathcal{F}$ of subgroups of G to play the part of the set $\mathcal{P}$ of p-subgroups of G and to observe that the Sylow p-subgroups are simply the maximal elements of $\mathcal{P}$. The right concept was formulated by Anderson ([1], 1975) in his successful attempt to localize to the subgroup lattice of a single group the theory of Fitting classes and injectors developed by Fischer, Gaschütz & Hartley a decade earlier. Anderson defined a <u>Fitting set</u> of a group G to be a G-invariant set $\mathcal{F}$ of subgroups satisfying

(FS1) <u>s_n-closure</u>: if K sn $H \in \mathcal{F}$, then $K \in \mathcal{F}$;

(FS2) <u>N_0-closure</u>: if $N_1, N_2 \trianglelefteq N = N_1N_2$ and $N_1, N_2 \in \mathcal{F}$, then $N \in \mathcal{F}$.

The set of p-subgroups of a group is clearly a Fitting set. More generally, if $\underline{F}$ is a Fitting class, then the set of $\underline{F}$-subgroups of G is obviously a Fitting set of G, although a group will usually have many Fitting sets not of this form. The main theorem for Fitting classes, and its proof, carry over, almost verbatim, to the Fitting set situation.

THEOREM 3.1 (Fischer, Gaschütz, Hartley and Anderson). <u>Let $\mathcal{F}$ be a Fitting set of a finite soluble group G. Then G has a unique conjugacy class of $\mathcal{F}$-injectors, namely subgroups H which satisfy</u>

(**) <u>if</u> K sn G, <u>then</u> $H \cap K$ <u>is</u> $\mathcal{F}$-<u>maximal in</u> K.

Thus Fitting sets play the part of p-subgroups in generalizing the Property (*) of Sylow p-subgroups. If $Inj_{\mathcal{F}}(G)$ denotes the conjugacy class of $\mathcal{F}$-injectors of G, we obtain

(***) if $H \in Inj_{\mathcal{F}}(G)$ and K sn G, then $H \cap K \in Inj_{\mathcal{F}}(K)$.

If $\mathcal{P}$ denotes the set of p-subgroups of G, we have $Inj_{\mathcal{P}}(G) = Syl_p(G)$. Therefore $\mathcal{F}$-injectors are exactly the generalizations of Sylow subgroups which capture their Property (*).

We now focus attention on the set Inj(G) of all injectors of a group G; thus $Inj(G) = \{ H \leq G \mid H \in Inj_{\mathcal{F}}(G)$ for some Fitting set $\mathcal{F} \}$. As an embedding property, $\mathcal{E}(G) = Inj(G)$ is remarkably well-behaved. If $\mathcal{F}$ is a Fitting set of G and $\phi: G \to \phi(G)$ an epimorphism, then $\phi(\mathcal{F})$ is a Fitting set of $\phi(G)$, and if $H \in In_{\mathcal{F}}(G)$, then $\phi(H)$ is a $\phi(\mathcal{F})$-injector of $\phi(G)$. In fact, results of Anderson show that more is true:

THEOREM 3.2. <u>The embedding property</u> Inj <u>is</u> Q-<u>invariant</u>.

Property (***) shows that if K sn G, then Inj(K) contains the set $\{ H \cap K \mid H \in Inj(G) \}$; however, these sets do not in general coincide (e.g. take $G = A_4$), and so Inj is not s_n-invariant. In contrast, the embedding property Inj^* defined by

$$Inj^*(G) = \{ H \leq G \mid H \in Inj_{\underline{\underline{F}}}(G) \text{ for some Fitting class } \underline{\underline{F}} \}$$

is indeed s_n-invariant (it would be interesting to know if Inj^* is the largest such contained in Inj), but this gain has to be set off against the fact that Inj^* is far from being Q-invariant. If H is an injector of G for a Fitting set (or a Fitting class) and if $H \leq L \leq G$, then H is an injector of L for the same Fitting set (class). Thus

THEOREM 3.3. <u>The embedding properties</u> Inj <u>and</u> Inj^* <u>are persistent</u>.

The subgroups in Inj(G) are defined indirectly in terms of the Fitting sets of G, but they can be characterized directly. We can prove

THEOREM 3.4. <u>Let</u> G <u>be a finite soluble group. Then</u> $H \in Inj(G)$ <u>if and only if the set of subnormal subgroups of the conjugates of</u> H <u>is</u> N_0-<u>closed</u>.

Injectors display many of the attractive properties of Sylow subgroups: they are pronormal, they satisfy the Frattini argument, and they either cover or avoid chief factors. There is a stronger version of pronormality defined as follows:

A subgroup H is <u>strongly pronormal</u>* in G if, for all primes p, a Sylow p-subgroup of H is a Sylow p-subgroup of some normal subgroup of G. (Write H spr G to denote this.)

In [3] Chambers showed that strongly pronormal subgroups are indeed pronormal. It is well known that if Σ is a Hall sytem of a finite soluble group G and if H is a pronormal subgroup of G, then Σ

---

*Called 'normally embedded' by many authors.

reduces into a unique conjugate of H. If $\underline{Pro}_{\Sigma}(G)$ ($\underline{Spro}_{\Sigma}(G)$) denotes the set of pronormal (strongly pronormal) subgroups of G into which Σ reduces, then $\underline{Pro}_{\Sigma}(G)$ is closed under intersections. Moreover, Fischer showed that the subset $\underline{Spro}_{\Sigma}(G)$ is actually a sublattice of $\mathcal{L}(G)$ and that the join operation is permutable product, i.e. $\langle H,L\rangle = HL$ for all $H,L \in \underline{Spro}_{\Sigma}(G)$. This suggests the following definition:

A subgroup embedding property $\mathcal{E}$ <u>admits a lattice</u> if for all groups G in the given universe

(1) $\mathcal{E}(G)$ contains a sublattice $\mathcal{H}(G)$ of $\mathcal{L}(G)$, and

(2) $\{ E^g \mid g \in G \} \cap \mathcal{H}(G) \neq \phi$ for all $E \in \mathcal{E}(G)$.

Normality, subnormality and strong pronormality are all examples of embedding properties which admit a lattice. It would be of interest to find others.

Finally, we return to the question of what information we need to identify the subgroups in $\mathcal{E}(G)$. Clearly pronormality can be determined from knowledge of $\mathcal{L}(G)$ and the conjugacy relations $\mathcal{C}_H(G)$, $H \leq G$. From its definition strong pronormality appears to require some additional arithmetical information, namely knowledge of the Sylow subgroups of the subgroups of G. However, Anderson proved the following theorem in [1].

THEOREM 3.5. <u>Any two of the following statements about a subgroup</u> H <u>of a finite soluble group</u> G <u>are equivalent</u>:

(a) H spr G;

(b) H <u>is an</u> $\mathcal{F}$-<u>injector of</u> G <u>for some subgroup-closed Fitting set</u> $\mathcal{F}$;

(c) <u>The set of subgroups of conjugates of</u> H <u>is</u> N_0-<u>closed</u>.

Since Criterion (c) can avidently be determined from knowledge of $\mathcal{L}(G)$ and $\mathcal{C}_H(G)$ $(H\leq G)$ alone, the embedding property "strong pronormality" is independent of knowledge of the Sylow subgroups of a group.

References

[1] ANDERSON, Wade: Injectors in Finite Solvable Groups, J. Algebra 36 (1975), 333-338.

[2] BEIDLEMAN, James C. Brewster, Ben, and Hauck, Peter: Fittingfunktoren in endlichen auflösbaren Gruppen, Math. Z. 182 (1983), 359-384.

[3] CHAMBERS, Graham A.: p-Normally Embedded Subgroups of Finite Soluble Groups, J. Algebra 16 (1970), 442-455.

SOLUBLE IRREDUCIBLE GROUPS OF AUTOMORPHISMS OF CERTAIN GROUPS OF CLASS TWO

Hermann Heineken, Mathematisches Institut
Universität Würzburg
Federal Republic of Germany

The following general result is due to Bryant and Kovács [1]: Given a finite group K operating faithfully on an elementary abelian noncyclic p-group T, there is a p-group P such that $P/\emptyset(P)$ is isomorphic to T, further $\mathrm{Aut}\, P/C_{\mathrm{Aut}\, P}(P/\emptyset(P))$ is isomorphic to K, and this quotient group operates on $P/\emptyset(P)$ "as" K on T. It is also known for p-groups of class 2, that every finite group K occurs as quotient group $\mathrm{Aut}\, P/\mathrm{Aut}_c P$ for some P (Heineken and Liebeck [4], U. Webb [8], Soules [7]). In most of these cases K operates semiregularly on a suitable basis of $P/\emptyset(P)$. On the other hand, some groups of nilpotency class two generated by three or four elements and their noncentral automorphism groups have been studied (Caranti [2], Daues and Heineken [3]).

In this note p-groups P (p odd) and soluble subgroups L of Aut P are considered satisfying the following conditions:

(I) $P/\emptyset(P)$ is of rank a prime q

(II) $P' = Z(P) = \Omega_1(P)$ is of rank $\binom{q}{2}$

(III) Every L-invariant proper subgroup of P is contained in P'.

Our main interest is in the quotient group $L/L \cap \mathrm{Aut}_c P = K$, which by (III) operates irreducibly on P/P'. We will see in Theorem 1 that there are only finitely many isomorphism classes left for this quotient group once q is fixed.

The results can be used to construct soluble complete groups, unlike those of Schuhmann [6] they have Fitting subgroups of nilpotency class two, but there are similarities in the search for operator isomorphic modules and for solutions of algebraic equations.

Before we prove the main result Theorem 1, we will provide the more numeric statements needed. Some examples will conclude this note.

Lemma 1: Assume that the element x of a field F satisfies simultaneously the following two equations:

$$x^q - 1 = 0 \quad , \quad \text{where } q \text{ is a prime, and}$$

$$1 + x - x^d = 0 \ .$$

Then there is a power $x^f = z$ such that

$$1 + z^a - z^b = 0 \quad \text{with} \quad 0 < a,b \leq [\sqrt{2q}] + 1 \ .$$

Proof:

We put $\quad w = [\sqrt{2q}] + 1$

and $\quad [\frac{dw}{q}] = s$.

If $d \leq w$, the statement of the lemma is true for $z=x$.

We assume now $d > w$ and so

(1) $s \geq 2$.

By definition of s, there are s natural numbers $a_1 < a_2 < \ldots < a_s \leq w$ such that

$$iq < a_i d < iq + d,$$

and the set

$$M = \{0, d , a_1 d-q , \ldots, a_i d-iq, \ldots, a_s d-sq\}$$

consists of $s+2$ pairwise different integers belonging to the interval

$$I = \{y \mid 0 \leq y \leq d\} \ .$$

We divide this interval I into $m = [\frac{s+2}{2}]$ subintervals J_k of equal length, that is

$$J_k = \{y \mid (k-1)\frac{d}{m} \leq y \leq k\frac{d}{m}\} \ .$$

If some $a_i d-iq$ belongs to J_1, we choose f such that $fa_i \equiv 1 \bmod q$ and hence $x = z^{a_i}$.

Now $\quad 1 + z^{a_i} - z^{a_i d-iq} = 0,$

and we have $a_i d-iq \leq \frac{d}{m} \leq \frac{2d}{s+1} \leq \frac{2q}{w} \leq w$.

If there is no $a_i d - iq$ belonging to J_1, the $s+1$ elements of M which are different from 0 are contained in the $m-1 = [\frac{s}{2}]$ remaining subintervals, so there is a subinterval J_r which contains three different elements of M.

If, furthermore, there is a pair $a_i d-iq$, $a_j d-jq$ such that $a_i - a_j$ and $(a_i d-iq)-(a_j d-jq)$ are positive, we choose f such that

$$f(a_i - a_j) \equiv 1 \bmod q$$

and go on as in the first case. If, one the other hand, no such pair among the three elements of J_r

$$a_i d-iq \ , \ a_j d-jq \ , \ a_k d-kq$$

exists and the three elements are given in their order of magnitude, then

$$2w \geq w + \frac{d}{m} \geq |(a_k d-kq)-(a_i d-iq)| + |a_k - a_i| =$$

$$(|(a_k d-kq)-(a_j d-jq)| + |a_k - a_j|)+(|a_j d-jq)-a_i d-iq)|+|a_j - a_i|)$$

One of the two last summands must be smaller than w. Assume first that the first one is, and choose f such that $f(a_k - a_j) \equiv 1$

Then $1 + z^{a_k - a_j} - z^{(a_k - a_j)d-(k-j)q} = 0$

and multiplication by $z^{a_j - a_k}$ yields

$$1 + z^{a_j - a_k} - z^{(a_k - a_j)d-(k-j)q+a_j-a_k} = 0 \ ,$$

a polynomial of the desired form.

The argument is analogous if the second summand is smaller than w. This concludes the proof.

<u>Lemma 2:</u> Let the element x of the field F satisfy simultaneously the two equations

$$x^q - 1 = 0$$

and

$$1 + x - x^d = 0.$$

Then

(i) char (F) $\leq 3 \cdot 2^{q-2}$.

If furthermore x is not contained in a proper subfield of F, then

(ii) F is of odd degree over its prime field except if q=3 and $|F| = 4$,

(iii) The degree of F over its prime field does not exceed $[\sqrt{2q}] + 1$.

<u>Proof:</u> (i) If x satisfies the two equations simultaneously, then char F divides

$$\det(E + X - X^d) ,$$

where E is the q x q identity matrix and X is the cyclic permutation of order q. Denote $E + X - X^d$ by M, we want to bound $|\det M|$. If A is some square matrix and $(t_1, t_2, \dots, t_m)$ some line of A with S_i the submatrix of A avoiding the given line and the ith collumn, then

$$|\det A| \leq \sum_i |t_i||S_i| < (\Sigma |t_i|)\operatorname{Max}_i |S_i| .$$

We use this formula for the first line of M and obtain

$$|\det M| \leq 3 \operatorname{Max} |S_i|$$

In each of the matrices S_i there is (at least) one line with only two non-zero entries, and this is also true for all other (proper) square submatrices. So, using the formula again, we have

$$|\det M| \leq 3 \cdot 2^{q-2} \cdot 1$$

(the last entry being completely fixed).
This proves (i).
For (ii), assume that F is constructed by adjoining x to the prime field F_p. If $[F:F_p]$ is even, there is a unique subfield L of F such that $[F:L] = 2$, and q is a divisor of $1 + |L| = 1+p^k$. There is some a in L such that

$$x^2+ax+1 = 0,$$

and $$-a = x + x^{p^k}$$

Now $$1 = x^d (x^{p^k})^d = (1+x)(1+x^{p^k}) = 1 - a + 1,$$

which yields $a = 1$ and $L = F_p$; so $q = 3$ and $p = 2$.

This shows (ii).
Statement (iii) is a direct consequence of Lemma 1.

<u>Theorem:</u> Let q be a prime and L a soluble group of automorphisms of the p-group P satisfying the conditions (I), (II), and (III). If $K = L/L \cap \operatorname{Aut}_c(P)$ we have for fixed q independent of p:

(a) K belongs to one of only finitely manny isomorphy classes,

(b) $K''' = 1$,

(c) if $K'' \neq 1$ then $Z(K/K'') = 1$ and $|K'/K''| = q$,
(d) K is cyclic or $Z(K) = 1$,
(e) if K is metacyclic, $K/K' = q$,
(f) the exponent of Fit(K) is bounded,
(g) Fit(K) is abelian,
(h) the rank of Fit(K) does not exceed $[\sqrt{2q}] + 1$.

Proof: K operates as an irreducible group of automorphisms on P/P', by (III), so Z(K) is cyclic. If $1 \neq Z(K) \neq K$, we find that Z(K) can not act irreducibly on P/P', otherwise it would generate the full ring of endomorphisms induced by K in P/P' which by the noncommutativity of K is not true. If $y \neq 1$ is an element of Z(K) it will therefore induce a power automorphism $a \to a^n$ on P/P' and P^p and a different power automorphism $b \to b^{n^2}$ on P'. This is a contradiction showing (d).
Consider now a sylow subgroup S of Fit(K). If S is cyclic, there is an irreducible S-module P^p in P' which is operator-isomorphic to P/P', and this is true only if

$$p^q - 1 \equiv 1 + p^r - p^s \equiv 0 \mod |S| .$$

On the other hand, if S is not cyclic, it does not operate irreducibly on P/P', and there is an element yP' such that $\langle yP' \rangle$ is S-invariant. There are q such S-invariant subgroups of P/P' and they are cyclically permuted by some element x of order q. Choose x such that $[y,x^{-1}]$ is occurring in the presentation of some element in P^p, where P itself is represented as generated by y and its $\langle x \rangle$-conjugates. Now the corresponding element in P^p generates an S-invariant subgroup and must therefore be equal to $\langle (x^{-d} y x^d)^p \rangle$ for suitable d. The operation of x on the linear forms of S now yields

$$1 + x = x^d ,$$

and we have furthermore $\quad x^q - 1 = 0.$

To find the exponent of Fit(K), we are left in both cases to find all numbers t such that

$$x^q - 1 \equiv 1 + x - x^d \equiv 0 \mod t.$$

As pointed out in the proof of Lemma 2(i), t is the divisor of some number smaller than $3 \cdot 2^{q-2}$.
The subgroup T of S centralizing yP' is normal in S, also S/T is cyclic.
Since $\bigcap_i T^{x^i} = 1$, we have that S is abelian.

So Fit(K) is abelian and of exponent bounded by a function of q. This shows (g) and (f). On the other hand K/Fit(K) operates as transitive permutation group on the q subgroups $\langle x^{-1}yx^{i}P'\rangle$ of P/P', if Fit(K) is noncyclic. In this case K/Fit(K) is either cyclic of order q or metacyclic with trivial centre and commutator subgroup of order q. If Fit(K) is cyclic, K/Fit(K) is trivial or of order q. This shows (e),(c) and (b). Finally (h) follows from Lemma 1, and all the statements (b)-(h) together yield (a).

Examples

Example 1. If $q > 3$ and $2q+1$ is a prime, then the metacyclic group of order $q(2q+1)$ is a quotient group L as laid down in Theorem 1.

Proof: Obviously $2q+1 \equiv 1 \mod 12$, and so 3 is a square mod $2q+1$. But then $3+1$ is some power of 3 mod $2q+1$. The prime p must be chosen as a square mod $2q+1$.

Example 2. If t is some prime and the prime q statisfies some equation

$$q(t^{3^m}-1) = t^{3^{m+1}}-1 ,$$

then the nonabelian extension of an elementary abelian group of order $t^{3^{m+1}}$ by the cyclic group of order q is group of automorphisms of some p-group of rank q.

Proof: Assume that F is the field of order $t^{3^{m+1}}$ and L is its unique subfield of order t^{3^m}. If a is some element of order q, it is a zero of some polynomial

$$x^3+ rx^2 +sx-1,$$

where r and s are elements of L. If $a + 1$ is also of order q, then (as in Lemma 2 (ii))

$$1 -r +s +1 = 1,$$

so $r = s+1$. We are looking for a polynomial

$$x^3 + (s+1)x^2 + sx \quad -1$$

which is irreducible over L. This means that the plynomial should not have a zero in L, and for all $x \in L\setminus\{0,-1\}$ we have

$$s \neq (x^2+x)^{-1}(x^3+x-1).$$

Since these are less inequalities than elements in L, s can be chosen to satisfy all these inequalities. Now we choose the metabelian group in such a way that conjugation by the element of order q operates like multiplication by a in F. The prime p must be congruent to 1 mod t.

Example 3. If $q = 2^r - 1$ is Mersenne prime, the metabelian group of order $q2^r$ is group of automorphisms of some p-group.

Proof: Since here all elements outside the prime field are of order q, the argument is comparatively simple and left to the reader. The prime p may be any odd prime.

The method mentioned in example 1 can be used for instance for
q = 11, 23, 29, 41 , ...
example 2 can be applied for q = 13 (for $t^{3^m} = 3$), 31(5), 73(8), 757 (27), 1723(41), ...

Remark 1: If $\frac{q-1}{2}$ is a prime bigger than 3, the soluble irreducible groups of automorphisms of a p-group of rank q are all cyclic or metacyclic. This follows from Theorem 1(h).

Remark 2: If q=3, the only groups possible are A_4 and S_4. Consulting the list in [3] the reader will observe that these groups are indeed never isomorphic to Aut P/Aut_cP (which in this case is $O_3(p)$).

Remark 3: Let R be a field of order p^q, and let t such that

$$1+p^r-p^s \equiv p^q-1 \equiv 0 \not\equiv p-1 \mod t$$

for some integers r and s.
The group M of matrices described by

$$M = \begin{pmatrix} 1 & a & b \\ 0 & 1 & a^{p^r} \\ 0 & 0 & 1 \end{pmatrix}, \quad a,b \in R$$

is a q-generated nilpotent p-group of class 2 and exponent p, further M possesses (see [4;Satz 5(2)]) an automorphism α of order t and there is an isomorphism γ of M/M' onto M' such that

$$\alpha\gamma = \gamma\alpha .$$

Denote by F the free group of rank q; we are now interested in the quotient group $W = F/F^{p^2}(F')^p F_3$. Obviously there is an isomorphism of M/M' onto $W/W'W^p$, we will call it σ, and there is an automorphism β of order t of W such that

$$(xM')^{\alpha\sigma} = (xM')^{\sigma\beta}.$$

We can find a β-invariant normal subgroup S such that

$$W^p \subseteq S \subset W'W^p$$

and $$W/S \cong M ,$$

and since t is prime to p, there is a β-invariant complement C of $S \cap W'$ in W'. By construction C is isomorphic to W^p, and this isomorphism ρ can be chosen such that

$$\beta\rho = \rho\beta$$

Now let $T = \{x(x^{\rho})^{-1} | x \in C\}$, then W/T is a group satisfying (I),(II,) (III) and admitting a cyclic group of automorphisms of order t (which then is irreducible in the sense of the title). This accounts for the cyclic groups L occurring in Theorem 1; it is possible to take care of the metacyclic ones accordingly. The constructions in the non-metacyclic case seem more transparent and are left to the reader.

References

[1] R.M. Bryant and L. Kovacs, Lie representations and groups of prime power order. J. London Math.Soc. (2) 17, 415-421 (1978)

[2] A. Caranti Automorphism groups of p-groups of class 2 and exponent p^2: A classification on 4 generators Annali di Mat 134, 93-146 (1983)

[3] G. Daues and H. Heineken Dualitäten und Gruppen der Ordnung p^6 Geometriae Dedicata 4, 215-220 (1975)

[4] H. Heineken and H. Liebeck The occurrence of finite groups in the automorphism group of nilpotent p-groups of class 2 Arch. Math. 25, 8-16 (1974)

[5] H. Heineken Gruppen mit kleinen abelschen Untergruppen Arch. Math. 29, 20-31 (1977)

[6] B. Schuhmann Über gewisse vollständige Gruppen ungerader Ordnung Dissertation Würzburg 1982

[7] P. Soules Construction of finite p-groups with prescribed group of non-central automorphisms to appear in Rend.Sem.Mat.U. Padova

[8] U.H.M. Webb The occurrence of groups as automorphisms of nilpotent p-groups Arch. Math. 37, 481-498 (1981)

ON AUTOMORPHISM GROUPS WHICH NORMALIZE AN ABELIAN NORMAL SUBGROUP

Hartmut Laue
Mathematisches Seminar der Universität
Olshausenstr. 40, Haus S12a
D 2300 Kiel 1
Bundesrepublik Deutschland

Dedicated to Professor B. Huppert on the occasion of his 60th birthday

An automorphism of a group G is called central if it fixes each coset of the center Z(G). The group $\mathrm{Aut}_c(G)$ of central automorphisms of G plays an important role in numerous group theoretic investigations. Its structure has been described, under certain finiteness conditions, as early as 1937, by Fitting [1]. For a finite group G it looks like that of the automorphism group of a finite abelian group, i.e., $\mathrm{Aut}_c(G)$ has a nilpotent normal subgroup N such that $\mathrm{Aut}_c(G)/N$ is a direct product of general linear groups over fields. On the other hand, Heineken and Liebeck [2] have shown that $\mathrm{Aut}(G)/\mathrm{Aut}_c(G)$ can be an arbitrary finite group. A result which seems to have escaped observation in this connection is the following:

Theorem I. Let G be a group with finite center Z(G). Then $\mathrm{Aut}_c(G)$ has a normal supplement W in Aut(G) such that $\mathrm{Aut}_c(G) \cap W$ is nilpotent. (In fact, $\mathrm{Aut}_c(G)/N$ is a direct factor of Aut(G)/N.)

It turns out that this theorem is a coarse special case of a much more general result which deals with groups of automorphisms normalizing an arbitrary abelian normal subgroup (instead of Z(G)). Moreover, the latter may be extended to rings, so that we prove it in a version containing applications for both groups and rings (Theorem 2). An analogue of Theorem I for algebras over fields is

Theorem II. Let G be a (not necessarily associative) algebra over a field F. Put A := Ann(G), C := $C_{\mathrm{Aut}(G)}(G/A)$, and suppose that $\dim_F A$ is finite. Then C has a normal supplement W in Aut(G) such that $W \cap C$ is nilpotent.

The main method in our proof is to exploit the results of [3] to represent automorphisms which normalize an abelian normal subgroup by certain 2×2 matrices. This matrix representation originates from a special semidirect decomposition of G and might be of interest on its own (Theorem 1).

We study properties of the entries of our 2×2 matrices to define a homomorphism of the automorphism group in question whose kernel is the desired supplement W.

Projection matrices

Let $(G,\square)$ be a group, A an abelian normal subgroup of G, and G_1 a subgroup of G which has a normal complement G_2 in G such that $G_2 \leq A$. We write π_j for the projection of G onto G_j with respect to the semidirect decomposition (G_1,G_2). If α is a mapping of G into G, then $\alpha_{ij} := \alpha_{|G_i}\pi_j$ is a mapping of G_i into G_j. The 2×2 matrix

$$\mathfrak{p}(\alpha) := (\alpha_{ij})$$

is called the projection matrix of α with respect to (G_1,G_2). Conversely, if α_{ij} is a mapping of G_i into G_j $(1 \leq i,j \leq 2)$, then we define the associate composite mapping $\mathfrak{c}(\alpha_{ij})$ of G into G by

$$(x_1\square x_2)^{\mathfrak{c}(\alpha_{ij})} := x_1^{\alpha_{11}}\square x_2^{\alpha_{21}}\square x_1^{\alpha_{12}}\square x_2^{\alpha_{22}}$$

for all $x_1 \in G_1$, $x_2 \in G_2$. We observe

(1) If α is a mapping of G into G such that $A^\alpha \subseteq A$ and $(x_1\square x_2)^\alpha = x_1^\alpha\square x_2^\alpha$ for all $x_1 \in G_1$, $x_2 \in G_2$, then $\alpha = \mathfrak{c}(\mathfrak{p}(\alpha))$.

(2) If, for $1 \leq i,j \leq 2$, α_{ij} is a mapping of G_i into G_j such that $1^{\alpha_{ij}} = 1$, then $\mathfrak{p}(\mathfrak{c}(\alpha_{ij})) = (\alpha_{ij})$.

If α is an endomorphism of $(G,\square)$ such that $A^\alpha \subseteq A$, then by comparison of $(x_1\square x_2\square y_1\square y_2)^\alpha$ and $(x_1\square x_2)^\alpha\square(y_1\square y_2)^\alpha$ (where $x_i, y_i \in G_i$) we get

(*) α_{ij} is a $\square$-homomorphism if $(i,j) \neq (1,2)$,
α_{12} is an α_{11}-cocycle of $(G_1,\square)$ into $(G_2,\square)$,

$$(x_2^{x_1})^{\alpha_{2j}} = (x_2^{\alpha_{2j}})^{x_1^{\alpha_{11}}} \quad \text{for all } x_i \in G_i \text{ and } j \in \{1,2\},$$

$$A_i^{\alpha_{ij}} \subseteq A_j \quad \text{for } 1 \leq i,j \leq 2,$$

where $A_1 := A \cap G_1$, $A_2 := G_2$ and an α_{11}-cocycle of $(G_1,\square)$ into $(G_2,\square)$ is a mapping f of G_1 into G_2 such that $(x_1 \square y_1)^f = x_1^{fy_1^{\alpha_{11}}} \square y_1^f$ for all $x_1, y_1 \in G_1$. By (1), $\alpha = \mathfrak{r}(\mathfrak{p}(\alpha))$. Conversely, if α_{ij} are mappings of G_i into G_j such that (*) holds, then $\mathfrak{r}(\alpha_{ij})$ is an endomorphism of G such that $A^\alpha \subseteq A$, and $\mathfrak{p}(\mathfrak{r}(\alpha_{ij})) = (\alpha_{ij})$, by (2).

We define the sum of two mappings $\alpha_{ij}, \beta_{ij} : G_i \to G_j$ by

$$x^{\alpha_{ij}+\beta_{ij}} := x^{\alpha_{ij}} \square x^{\beta_{ij}} \quad \text{for all } x \in G_i,$$

and the product of two projection matrices by formal matrix multiplication. Then, if α, β are endomorphisms of G such that $A^\alpha, A^\beta \subseteq A$, we have

$$(x_1 \square x_2)^{\alpha\beta} = x_1^{\alpha_{11}\beta_{11}} \square \prod_{\substack{1 \leq i,j,k \leq 2 \\ (i,j,k) \neq (1,1,1)}}^{\square} x_i^{\alpha_{ij}\beta_{jk}} = ((x_1 \square x_2)^{\mathfrak{r}(\mathfrak{p}(\alpha)\mathfrak{p}(\beta))}$$

for all $x_1 \in G_1$, $x_2 \in G_2$. Summarizing, this yields

Lemma 1. Let $(G,\square)$ be a group, A an abelian normal subgroup of G, (G_1,G_2) a semidirect decomposition of G such that the normal subgroup G_2 of G is contained in A. Then we have

(i) A mapping α of G into G is an $\square$-endomorphism with $A^\alpha \subseteq A$ if and only if the entries α_{ij} of its projection matrix $\mathfrak{p}(\alpha)$ satisfy (*).

(ii) If α, β are endomorphisms of $(G,\square)$ and $A^\alpha, A^\beta \subseteq A$, then $\mathfrak{p}(\alpha\beta) = \mathfrak{p}(\alpha)\mathfrak{p}(\beta)$.

Furthermore, it is clear that the mapping $\alpha \to \mathfrak{p}(\alpha)$ (α as in (ii)) is injective.

Now suppose that we have a further composition (a "multiplication") on G such that both distributive laws hold:

$$(x \square y)z = xz \square yz, \quad z(x \square y) = zx \square zy$$

for all $x,y,z \in G$. Then G satisfies all axioms of a (not necessarily associative) ring except the commutative law of addition. Such a structure is called a skew ring ([5],[4]). An ideal of G is defined to be a normal subgroup H of $(G,\square)$ such that $GH, HG \subseteq H$. If, moreover, $HH = 0$, H is called a zero ideal. A subgroup of $(G,\square)$ which is closed under the skew ring multiplication is called a sub-skew ring. A semidirect decomposition of the skew ring G is a pair (G_1,G_2) where G_2 is an ideal, G_1 a sub-skew ring of G such that $G_1 \square G_2 = G$, $G_1 \cap G_2 = 1$. We first wish to extend the statement in Lemma 1(i) to skew rings: Let A be a zero ideal of G such that $(A,\square)$ is commutative, and (G_1,G_2) a semidirect decomposition of G such that $G_2 \subseteq A$. Let α be an endomorphism of the group $(G,\square)$ such that $A^\alpha \subseteq A$. Then we compare $((x_1 \square x_2)(y_1 \square y_2))^\alpha$ and $(x_1 \square x_2)^\alpha (y_1 \square y_2)^\alpha$ (where $x_i, y_i \in G_i$) and find that α is a multiplicative homomorphism if and only if the following conditions hold:

(**) α_{11} is a multiplicative homomorphism,
α_{12} is an α_{11}-derivation of G_1 into G_2,
$(x_1x_2)^{\alpha_{2j}} = x_1^{\alpha_{11}} x_2^{\alpha_{2j}}$, $(x_2x_1)^{\alpha_{2j}} = x_2^{\alpha_{2j}} x_1^{\alpha_{11}}$ for all $x_i \in G_i$ and $j \in \{1,2\}$,

where an α_{11}-derivation of G_1 into G_2 is an α_{11}-cocycle f of $(G_1,\square)$ into $(G_2,\square)$ such that $(x_1y_1)^f = x_1^{\alpha_{11}} y_1^f \square x_1^f y_1^{\alpha_{11}}$. Therefore, we have the following skew ring version of Lemma 1(i):

<u>Lemma 2</u>. Let G be a skew ring, A a zero ideal of G such that $(A,\square)$ is abelian, and (G_1,G_2) a semidirect decomposition of G such that $G_2 \subseteq A$. Then a mapping α of G into G is an endomorphism with $A^\alpha \subseteq A$ if and only if the entries α_{ij} of its projection matrix $\mathfrak{p}(\alpha)$ satisfy (*) and (**).

As before, let A be a zero ideal of a skew ring G such that $(A,\square)$ is abelian. Let Ω be a set of endomorphisms of certain subgroups of $(G,\square)$ which contain A; i.e., each $\omega \in \Omega$ is an endomorphism of some subgroup U_ω of $(G,\square)$ such that $A \leq U_\omega$. Suppose that Ω contains all endomorphisms of $(A,\square)$ which are induced by G per right and left multiplication and conjugation. We assume that A satisfies both chain conditions for Ω-subgroups. Here, as usual, an Ω-subgroup of A is a subgroup B of $(A,\square)$ such that $B^\omega \subseteq B$ for all $\omega \in \Omega$. Any such B is an ideal of G. We shall need the well-known fact that

(3) $J(\mathrm{End}_\Omega(B,\square))$ is nilpotent, and $\mathrm{End}_\Omega(B,\square)/J(\mathrm{End}_\Omega(B,\square))$ is a direct sum of finitely many full matrix rings over skew fields.

(Here, the functor J denotes the Jacobson radical of a ring.) An arbitrary subgroup H of $(G,\square)$ need not be contained in U_ω for $\omega \in \Omega$. Therefore we modify the familiar notion of an Ω-subgroup of G: A subgroup H of $(G,\square)$ is called an <u>Ω-subgroup</u> if $(H \cap U_\omega)^\omega \subseteq H \cap U_\omega$ for all $\omega \in \Omega$. A semidirect decomposition (G_1,G_2) of G is called a <u>semidirect Ω-decomposition</u> if G_1, G_2 are Ω-subgroups of G. Then, if $G_2 \subseteq A$, we have

$$\mathfrak{p}(\omega) = \begin{pmatrix} \omega_{11} & 0 \\ 0 & \omega_{22} \end{pmatrix}$$ where $\mathfrak{p}(\omega)$ is formed with respect to the semidirect Ω-decomposition $(G_1 \cap U_\omega, G_2)$ of U_ω, and 0 is the trivial mapping $G_j \to 1$.

If δ is a mapping of some Ω-subgroup V of G into some Ω-subgroup W of G, we say that <u>δ commutes with Ω</u> if $(V \cap U_\omega)^\delta \subseteq W \cap U_\omega$ and $x^{\omega\delta} = x^{\delta\omega}$ for all $x \in V \cap U_\omega$ and for all $\omega \in \Omega$.

If α is an injective endomorphism of $(G,\square)$ and ω is an endomorphism of a subgroup U of $(G,\square)$, then $\omega^\alpha := \alpha^{-1}\omega\alpha$ is an endomorphism of U^α. Hence $\mathrm{Aut}(G,\square)$, and a fortiori $\mathrm{Aut}(G)$, act on the set of all endomorphisms of all subgroups of $(G,\square)$. We put

$$D := \{ \alpha \mid \alpha \in N_{\mathrm{Aut}(G)}(A),\ \Omega^\alpha = \Omega \} .$$

Our next aim is to describe the elements of D via the entries of their projection matrices. Throughout the following we assume that

\# (G_1,G_2) is a semidirect Ω-decomposition of G such that G_2 is a
\# maximal element of $\{B \mid B$ is an Ω-subgroup of A which has an Ω-com-
\# plement in G$\}$.

As before, we put $A_1 := A \cap G_1$, $A_2 := G_2$. Moreover, we put $\Omega_A := \{\omega_{|A} \mid \omega \in \Omega\}$. Then $\Omega_A \subseteq \mathrm{End}(A)$.

As we have observed in [4], the results in [3, 3.] may easily be extended to skew rings with operator sets Ω: The set

$$R^{\Delta,\Omega}_{(G,A)} := \{f \mid f{:}G\to A,\ (x\square y)^f = x^f y \square y^f,\ (xy)^f = xy^f \square x^f y \text{ for all } x,y\in G,\ f \text{ commutes with } \Omega\}$$

is a ring with respect to the "addition" $\square$: $x^{f\square g} := x^f \square x^g$, and the "multiplication" $\cdot$: $x^{(f\cdot g)} := (x^f)^g$. D acts per conjugation on $R^{\Delta,\Omega}_{(G,A)}$, and the group $Q(R^{\Delta,\Omega}_{(G,A)})$ of all quasiregular elements of $R^{\Delta,\Omega}_{(G,A)}$ is D-isomorphic to $C_{\mathrm{Aut}(G)}(G/A,\Omega)$, the group of all automorphisms of G which centralize G/A and Ω. A canonical D-isomorphism of $C_{\mathrm{Aut}(G)}(G/A,\Omega)$ onto $Q(R^{\Delta,\Omega}_{(G,A)})$ is given by the mapping

(4) $\quad \bar{\ }: \alpha \to f_\alpha$ where $x^{f_\alpha} = [x,\alpha]$ for all $x \in G$

(cf. [3,3.1], [4,1.4]). As in [3] , we let N_J be the D-invariant nilpotent subgroup of $C_{\mathrm{Aut}(G)}(G/A,\Omega)$ such that

(5) $\quad \overline{N_J} = J(R^{\Delta,\Omega}_{(G,A)})$.

We shall need the "skew ring version" of [3,3.6] which implies that

(6) $R^{\Delta,\Omega}_{(G_1,A_1)}$ is nilpotent,

and the "skew ring version" of [3,(11)] which is

(7) If A has a complement K in G, then $\mathrm{Ann}_R(K)$ is a complement of $\mathrm{Ann}_R(A)$ in R and isomorphic to $\mathrm{End}_\Omega(A)$.

Here, $R := R^{\Delta,\Omega}_{(G,A)}$ and $\mathrm{Ann}_R(U) := \{ f \mid f \in R,\ U^f = 1 \}$ for all $U \subseteq G$.

<u>Lemma 3</u>. Let $\alpha, \beta, \gamma \in D$, $j,k,l \in \{1,2\}$. Then

(i) $x^{\omega\alpha_{jk}} = x^{\alpha_{jk}\omega^\alpha}$ for all $\omega \in \Omega$, $x \in G_j \cap U_\omega$.

(ii) If $\alpha\beta, \gamma \in C_D(\Omega_A)$, then $\alpha_{jk|A_j}\gamma_{kl}\beta_{lj} \in \mathrm{End}_\Omega(A_j)$, and $\alpha_{2k}\gamma_{kl}\beta_{12} \in J(\mathrm{End}_\Omega(G_2))$ for $(k,l) \neq (2,2)$. In particular, $\alpha_{21}\beta_{12} \in J(\mathrm{End}_\Omega(G_2))$.

(iii) If $\beta = \alpha^{-1}$, then $\alpha_{22}\beta_{22} \equiv \mathrm{id}_{G_2} \mod J(\mathrm{End}_\Omega(G_2))$.

<u>Proof</u>. (i) If $x_1 \in G_1 \cap U_\omega$, $x_2 \in G_2$, then $x_1^{\omega\alpha_{11}} \square x_2^{\omega\alpha_{21}} \square x_1^{\omega\alpha_{12}} \square x_2^{\omega\alpha_{22}} = (x_1 \square x_2)^{\omega\alpha} = (x_1 \square x_2)^{\alpha\omega^\alpha} = x_1^{\alpha_{11}\omega^\alpha} \square x_2^{\alpha_{21}\omega^\alpha} \square x_1^{\alpha_{12}\omega^\alpha} \square x_2^{\alpha_{22}\omega^\alpha}$. Putting $x_1 = 1$, $x_2 = 1$ resp., and using the hypothesis that G_1, G_2 are Ω-subgroups, we obtain (i).

(ii) By (i), we have $\omega_{|A_j}\alpha_{jk}\gamma_{kl}\beta_{lj} = \alpha_{jk|A_j}\gamma_{kl}\beta_{lj}\omega^{\alpha\gamma\beta} = \alpha_{jk|A_j}\gamma_{kl}\beta_{lj}\omega$, since $C_D(\Omega_A)$ is a normal subgroup of D. Hence $\alpha_{jk}\gamma_{kl}\beta_{lj} \in \mathrm{End}_\Omega(A_j)$. Now let $\delta \in \mathrm{End}_\Omega(G_2)$. By (3) and [3,1.5] it suffices to show that $\alpha_{2j}\gamma_{kl}\beta_{12}\delta$ is nilpotent if $(k,l) \neq (2,2)$. Suppose first that $l = 1$. For $x_1, y_1 \in G_1$, we have, by Lemma 2,

$$\begin{aligned}(x_1 \square y_1)^{\beta_{12}\delta\alpha_{2k}\gamma_{k1}} &= (x_1^{\beta_{12}y_1^{\beta_{11}}} \square y_1^{\beta_{12}})^{\delta\alpha_{2k}\gamma_{k1}} \\ &= (x_1^{\beta_{12}\delta y_1^{\beta_{11}}} \square y_1^{\beta_{12}\delta})^{\alpha_{2k}\gamma_{k1}} \\ &= (x_1^{\beta_{12}\delta\alpha_{2k}y_1^{\beta_{11}\alpha_{11}}} \square y_1^{\beta_{12}\delta\alpha_{2k}})^{\gamma_{k1}} \\ &= x_1^{\beta_{12}\delta\alpha_{2k}\gamma_{k1}y_1^{\beta\alpha}} \square y_1^{\beta_{12}\delta\alpha_{2k}\gamma_{k1}},\end{aligned}$$

and

$$\begin{aligned}(x_1y_1)^{\beta_{12}\delta\alpha_{2k}\gamma_{k1}} &= (x_1^{\beta_{11}}y_1^{\beta_{12}}\square x_1^{\beta_{12}}y_1^{\beta_{11}})^{\delta\alpha_{2k}\gamma_{k1}} \\ &= (x_1^{\beta_{11}}y_1^{\beta_{12}\delta}\square x_1^{\beta_{12}\delta}y_1^{\beta_{11}})^{\alpha_{2k}\gamma_{k1}} \\ &= (x_1^{\beta_{11}\alpha_{11}}y_1^{\beta_{12}\delta\alpha_{2k}}\square x_1^{\beta_{12}\delta\alpha_{2k}}y_1^{\beta_{11}\alpha_{11}})^{\gamma_{k1}} \\ &= x_1^{\beta\alpha}y_1^{\beta_{12}\delta\alpha_{2k}\gamma_{k1}}\square x_1^{\beta_{12}\delta\alpha_{2k}\gamma_{k1}}y_1^{\beta\alpha} ,\end{aligned}$$

finally for $\omega \in \Omega$ and $x_1 \in G_1 \cap U_\omega$

$$x_1^{\omega\beta_{12}\delta\alpha_{2k}\gamma_{k1}} = x_1^{\beta_{12}\delta\alpha_{2k}\gamma_{k1}\omega^{\beta\alpha\gamma}} .$$

As $\alpha\beta,\ \gamma \in C_D(\Omega_{|A}) \trianglelefteq D$, we also have $\beta\alpha,\ \beta\alpha\gamma \in D$. Hence our equations show that $\beta_{12}\delta\alpha_{2k}\gamma_{k1} \in R^{\Delta,\Omega}_{(G_1,A_1)}$. This ring is nilpotent, by (6). Therefore, the equation $(\alpha_{2k}\gamma_{k1}\beta_{12}\delta)^{n+1} = \alpha_{2k}\gamma_{k1}(\beta_{12}\delta\alpha_{2k}\gamma_{k1})^n\beta_{12}\delta$ implies our claim. The case $l=2$, $k=1$ may be treated similarly, by considering the mapping $\gamma_{12}\beta_{22}\delta\alpha_{21}$.

(iii) follows from (ii), as $id_{G_2} = \alpha_{21}\beta_{12} + \alpha_{22}\beta_{22}$.

<u>Theorem 1</u>. Let α be an endomorphism of G such that $A^\alpha \subseteq A$. Then $\alpha \in D$ if and only if the following conditions hold:

(i) $\alpha_{11} \in Aut(G_1)$, $\alpha_{22} \in Aut(G_2)$,

(ii) $\alpha_{jk}\alpha_{kk}^{-1}$, $\alpha_{jj}^{-1}\alpha_{jk}$ commute with Ω, for all $j,k \in \{1,2\}$,

(iii) $$\mathfrak{p}(\Omega) = \left\{ \begin{pmatrix} \omega^{\alpha_{11}} & 0 \\ 0 & \omega^{\alpha_{22}} \end{pmatrix} \;\middle|\; \omega \in \Omega \right\} .$$

<u>Proof</u>. Let $\alpha \in D$. We put $(\beta_{ij}) := \mathfrak{p}(\alpha^{-1})$. Then $id_{G_1} = \alpha_{11}\beta_{11} + \alpha_{12}\beta_{21}$, hence $\alpha_{11}\beta_{11} = id_{G_1} - \alpha_{12}\beta_{21}$. By Lemma 2, $\alpha_{11}\beta_{11} \in End(G_1)$, and by Lemma 3(ii) and (3),

(8) $\alpha_{21}\beta_{12}$ is nilpotent.

Interchanging α and α^{-1} in this argument, we obtain

(9) $\beta_{21}\alpha_{12}$ is nilpotent.

Now the identity $(\alpha_{12}\beta_{21})^{n+1} = \alpha_{12}(\beta_{21}\alpha_{12})^n\beta_{21}$ shows that $\alpha_{12}\beta_{21}$ is nilpotent, too. Hence $\alpha_{11}\beta_{11}$ is an automorphism of G_1. Again by interchanging α and α^{-1} we see that $\beta_{11}\alpha_{11}$ also is an automorphism of G_1.

Therefore, α_{11} is an automorphism. Moreover, $id_{G_2} = \alpha_{21}\beta_{12} + \alpha_{22}\beta_{22}$ so that $\alpha_{22}\beta_{22}$ is an automorphism of G_2, by (8). As before, this also holds for $\beta_{22}\alpha_{22}$, whence α_{22} is an automorphism. By symmetry, β_{11} and β_{22} are automorphisms, and the equations $\alpha_{11}\beta_{12} + \alpha_{12}\beta_{22} = 0$, $\alpha_{21}\beta_{11} + \alpha_{22}\beta_{21} = 0$ yield

$$(10) \qquad \alpha_{11}^{-1}\alpha_{12} = -\beta_{12}\beta_{22}^{-1}\,, \quad \alpha_{22}^{-1}\alpha_{21} = -\beta_{21}\beta_{11}^{-1}\,.$$

Let $\omega \in \Omega$. For all $x \in G_1 \cap U_{\omega^\alpha}$ we have

$$\begin{aligned} x^{\omega^\beta} &= (x^{\alpha_{11}\omega} \square x^{\alpha_{12}\omega})^\beta \\ &= x^{\alpha_{11}\omega\beta_{11}} \square x^{\alpha_{12}\omega\beta_{21}} \square x^{\alpha_{11}\omega\beta_{12}} \square x^{\alpha_{12}\omega\beta_{22}}\,, \end{aligned}$$

hence $x^{\alpha_{11}\omega\beta_{12}} \square x^{\alpha_{12}\omega\beta_{22}} = 1$, i.e., $y^{\omega\beta_{12}\beta_{22}^{-1}} = y^{-\alpha_{11}^{-1}\alpha_{12}\omega}$ for all $y \in G_1 \cap U_\omega$. Now (10) implies that $\alpha_{11}^{-1}\alpha_{12}$ commutes with Ω. If we let $x \in G_2$ and argue analoguosly, we see that $\alpha_{22}^{-1}\alpha_{21}$ commutes with Ω, too. As this also holds for β instead of α, a further application of (10) yields (ii).
Finally, (iii) is a trivial consequence of the identity

$$\not p(\omega^\alpha) = \begin{pmatrix} \omega_1^{\alpha_{11}} & 0 \\ 0 & \omega_2^{\alpha_{22}} \end{pmatrix}\,.$$

Conversely, suppose that (i), (ii), (iii) hold. Let $x_1, y_1 \in G_1$. By Lemma 2, we have

$$\begin{aligned} (x_1 \square y_1)^{\alpha_{11}^{-1}\alpha_{12}\alpha_{22}^{-1}\alpha_{21}} &= (x_1^{\alpha_{11}^{-1}\alpha_{12}y_1} \square y_1^{\alpha_{11}^{-1}\alpha_{12}})^{\alpha_{22}^{-1}\alpha_{21}} \\ &= (x_1^{\alpha_{11}^{-1}\alpha_{12}\alpha_{22}^{-1}y_1^{\alpha_{11}^{-1}}} \square y_1^{\alpha_{11}^{-1}\alpha_{12}\alpha_{22}^{-1}})^{\alpha_{21}} \\ &= x_1^{\alpha_{11}^{-1}\alpha_{12}\alpha_{22}^{-1}\alpha_{21}y_1} \square y_1^{\alpha_{11}^{-1}\alpha_{12}\alpha_{22}^{-1}\alpha_{21}}\,, \end{aligned}$$

$$\begin{aligned} (x_1 y_1)^{\alpha_{11}^{-1}\alpha_{12}\alpha_{22}^{-1}\alpha_{21}} &= (x_1^{\alpha_{11}^{-1}} y_1^{\alpha_{11}^{-1}})^{\alpha_{12}\alpha_{22}^{-1}\alpha_{21}} \\ &= (x_1 y_1^{\alpha_{11}^{-1}\alpha_{12}} \square x_1^{\alpha_{11}^{-1}\alpha_{12}} y_1)^{\alpha_{22}^{-1}\alpha_{21}} \\ &= (x_1^{\alpha_{11}^{-1}} y_1^{\alpha_{11}^{-1}\alpha_{12}\alpha_{22}^{-1}} \square x_1^{\alpha_{11}^{-1}\alpha_{12}\alpha_{22}^{-1}} y_1^{\alpha_{11}^{-1}})^{\alpha_{21}} \end{aligned}$$

$$= x_1 y_1^{\alpha_{11}^{-1}\alpha_{12}\alpha_{22}^{-1}\alpha_{21}} \square x_1^{\alpha_{11}^{-1}\alpha_{12}\alpha_{22}^{-1}\alpha_{21}} y_1 ,$$

and, by (ii), $\alpha_{11}^{-1}\alpha_{12}\alpha_{22}^{-1}\alpha_{21}$ commutes with Ω. Therefore, $\alpha_{11}^{-1}\alpha_{12}\alpha_{22}^{-1}\alpha_{21}$ is an element of $R^{\Delta,\Omega}_{(G_1,A_1)}$, hence nilpotent, by (6). This implies

(11) $\alpha_{21}\alpha_{11}^{-1}\alpha_{12}\alpha_{22}^{-1}$ is a nilpotent endomorphism of G_2.

Now, if $x_1 \in G_1$, $x_2 \in G_2$ such that $x_1 \square x_2 \in \ker \alpha$, we have $x_1^{\alpha_{11}} \square x_2^{\alpha_{21}} = 1$ $= x_1^{\alpha_{12}} \square x_2^{\alpha_{22}}$, hence $x_1 = x_2^{-\alpha_{21}\alpha_{11}^{-1}}$ and $x_2^{-\alpha_{21}\alpha_{11}^{-1}\alpha_{12}} \square x_2^{\alpha_{22}} = 1$, i.e., $x_2 =$ $x_2^{\alpha_{21}\alpha_{11}^{-1}\alpha_{12}\alpha_{22}^{-1}}$. By (11), this means that $x_2 = 1$, which implies that $x_1 = 1$, too. This shows that α is injective. Therefore, if $\omega \in \Omega$, then $\omega^\alpha :=$ $\alpha^{-1}\omega\alpha$ is an endomorphism of $(U_\omega)^\alpha$. We prove that $\Omega^\alpha = \Omega$: If $\omega, \omega^* \in \Omega$ such that $\mathfrak{p}(\omega^*) = \begin{pmatrix} \omega^{\alpha_{11}} & 0 \\ 0 & \omega^{\alpha_{22}} \end{pmatrix}$, then $U_{\omega^*} \cap G_1 = (U_\omega \cap G_1)^{\alpha_{11}}$ and therefore $U_{\omega^*} = A \square (U_\omega \cap G_1)^{\alpha_{11}} = (U_\omega)^\alpha$. Let $x_1 \in G_1$, $x_2 \in G_2$ such that $x_1 \square x_2 \in U_\omega$. Then, by (ii),

$$\begin{aligned}(x_1 \square x_2)^{\omega\alpha} &= x_1^{\omega\alpha_{11}} \square x_2^{\omega\alpha_{21}} \square x_1^{\omega\alpha_{12}} \square x_2^{\omega\alpha_{22}} \\ &= x_1^{\omega\alpha_{11}} \square x_2^{\alpha_{21}\omega^{\alpha_{11}}} \square x_1^{\alpha_{12}\omega^{\alpha_{22}}} \square x_2^{\omega\alpha_{22}} \\ &= (x_1 \square x_2)^{\alpha\omega^*}.\end{aligned}$$

Thus $\omega^* = \omega^\alpha$. We conclude that $\Omega^\alpha = \Omega$: If $\omega \in \Omega$, then, by (iii), there exists an element $\omega^* \in \Omega$ such that $\mathfrak{p}(\omega^*) = \begin{pmatrix} \omega^{\alpha_{11}} & 0 \\ 0 & \omega^{\alpha_{22}} \end{pmatrix}$. Hence $\omega^\alpha =$ $\omega^* \in \Omega$. Therefore, $\Omega^\alpha \subseteq \Omega$. Moreover, we may find an element $\bar{\omega} \in \Omega$ such that $\mathfrak{p}(\omega) = \begin{pmatrix} \bar{\omega}^{\alpha_{11}} & 0 \\ 0 & \bar{\omega}^{\alpha_{22}} \end{pmatrix}$, by (iii). Hence $\omega = \bar{\omega}^\alpha$, yielding $\Omega \subseteq \Omega^\alpha$. The equality $\Omega = \Omega^\alpha$ implies

(12) If X is an Ω-subgroup of A, so is X^α,

as $X^{\alpha\omega} = X^{\alpha(\omega^{\alpha^{-1}})^{\alpha}} = X^{(\omega^{\alpha^{-1}})\alpha} \subseteq X^{\alpha}$ for all $\omega \in \Omega$. Since α is injective and A satisfies the descending chain condition for Ω-subgroups, (12) implies that $A = A^{\alpha}$. We use this equality to show that α is onto: Let $x \in G$. By (1), we may find an element $y \in G_1$ such that $A\square y^{\alpha_{11}} = A\square x$. Then $y^{\alpha}\square x^{-1} \in A$. Hence $y^{\alpha}\square x^{-1} = (b^{-1})^{\alpha}$ for some suitable element $b \in A$. This implies $x = (b\square y)^{\alpha}$, and the proof of Theorem 1 is complete.

Applications

We next characterize the elements of $N_{\mathfrak{J}}$ (cf. (5)).

Lemma 4. Let $\alpha \in C_{\mathrm{Aut}(G)}(G/A,\Omega)$. Then $\alpha \in N_{\mathfrak{J}}$ if and only if $\alpha_{22} \equiv \mathrm{id}_{G_2}$ mod $\mathfrak{J}(\mathrm{End}_{\Omega}(G_2))$.

Proof. As $\alpha \in C_{\mathrm{Aut}(G)}(G/A,\Omega)$, we have $\alpha_{11} \in C_{\mathrm{Aut}(G_1)}(G_1/A_1)$, $\alpha_{22} \in \mathrm{Aut}(G_2)$ by Theorem 1, and we may define $f_{\alpha_{jj}} : G_j \to A_j$ for $j \in \{1,2\}$ by

$$\mathfrak{p}(\alpha) = \begin{pmatrix} \mathrm{id}_{G_1}+f_{\alpha_{11}} & \alpha_{12} \\ \alpha_{21} & \mathrm{id}_{G_2}+f_{\alpha_{22}} \end{pmatrix}.$$

$$\text{Then } \mathfrak{p}(f_{\alpha}) = \begin{pmatrix} f_{\alpha_{11}} & \alpha_{12} \\ \alpha_{21} & f_{\alpha_{22}} \end{pmatrix} = \begin{pmatrix} f_{\alpha_{11}} & 0 \\ \alpha_{21} & 0 \end{pmatrix} + \begin{pmatrix} 0 & \alpha_{12} \\ 0 & f_{\alpha_{22}} \end{pmatrix},$$

and $\mathfrak{r}\begin{pmatrix} f_{\alpha_{11}} & 0 \\ \alpha_{21} & 0 \end{pmatrix} \in R^{\Delta,\Omega}_{(G,A_1)} \subseteq \mathfrak{J}(R^{\Delta,\Omega}_{(G,A)})$ by [4,1.5(i),1.6(i)],

$g := \mathfrak{r}\begin{pmatrix} 0 & \alpha_{12} \\ 0 & f_{\alpha_{22}} \end{pmatrix} \in R^{\Delta,\Omega}_{(G,G_2)}$ by [4,1.5(i)]. Thus

(13) $\alpha \in N_{\mathfrak{J}}$ if and only if $g \in \mathfrak{J}(R^{\Delta,\Omega}_{(G,G_2)})$.

For all $h \in R^{\Delta,\Omega}_{(G,G_2)}$ we have $\mathfrak{p}((gh)^n) = (\mathfrak{p}(gh))^n = \begin{pmatrix} 0 & \alpha_{12}h_{22}(f_{\alpha_{22}}h_{22})^{n-1} \\ 0 & (f_{\alpha_{22}}h_{22})^n \end{pmatrix}$.

Hence, by (1), gh is nilpotent if and only if $f_{\alpha_{22}}h_{22}$ is nilpotent. By (7),

$\{h_{22} | h \in R^{\Delta,\Omega}_{(G,G_2)}\} = \mathrm{End}_\Omega(G_2)$. By means of [4,1.5], (3), and [3,3.2] (skew ring version) we conclude that $g \in J(R^{\Delta,\Omega}_{(G,G_2)})$ if and only if $f_{\alpha_{22}} \in J(\mathrm{End}_\Omega(G_2))$, i.e., if $\alpha_{22} \equiv \mathrm{id}_{G_2} \mod J(\mathrm{End}_\Omega(G_2))$. By virtue of (13), the proof is complete.

For all $\alpha \in D$, we put $\hat{\alpha} = \mathfrak{c}\begin{pmatrix} \mathrm{id}_{G_1} & \alpha_{12}\alpha_{22}^{-1} \\ \alpha_{21}\alpha_{11}^{-1} & \mathrm{id}_{G_2} \end{pmatrix}$, $\check{\alpha} = \mathfrak{c}\begin{pmatrix} \alpha_{11} & 0 \\ 0 & \alpha_{22} \end{pmatrix}$. By Theorem 1, $\hat{\alpha}$, $\check{\alpha} \in D$, and $\alpha = \hat{\alpha}\check{\alpha}$, by Lemma 1(ii). By Lemma 4, $\hat{\alpha} \in N_J$. Hence the stabilizer of the semidirect decomposition (G_1,G_2) in D is a supplement of N_J in D. This is a special case of [4,3.3]. If $\alpha \in C_{\mathrm{Aut}(G)}(G/A,\Omega)$, then $\check{\alpha} \in C_{\mathrm{Aut}(G)}(G/A,\Omega)$, too. After these preparations, we are able to prove

<u>Theorem 2</u>. Put $C := C_{\mathrm{Aut}(G)}(G/A,\Omega)$, $V := C_D(\Omega_A)$. Then V contains a D-invariant subgroup W such that $WC = V$, $W \cap C = N_J$.

<u>Proof</u>. We put $E := \mathrm{End}_\Omega(G_2)$. If $\alpha \in V$, then $\alpha_{22} \in E$. Let

$$\varphi: V \to E/J(E)$$
$$\alpha \to \alpha_{22} + J(E).$$

For all $\alpha, \beta \in V$ we have $(\alpha\beta)^\varphi = \alpha_{21}\beta_{12} + \alpha_{22}\beta_{22} + J(E) = \alpha_{22}\beta_{22} + J(E)$, by the last assertion of Lemma 3(ii). Thus φ is a homomorphism of V into the group of units of $E/J(E)$. We put $W := \ker\varphi$. Then $W \cap C = N_J$, by Lemma 4. If $\alpha \in V$, then $\check{\alpha} \in V$.

$$\alpha^{(1)} := \mathfrak{c}\begin{pmatrix} \alpha_{11} & 0 \\ 0 & \mathrm{id}_{G_2} \end{pmatrix} \quad , \quad \alpha^{(2)} := \mathfrak{c}\begin{pmatrix} \mathrm{id}_{G_1} & 0 \\ 0 & \alpha_{22} \end{pmatrix}$$

are elements of D. We have $\alpha^{(1)} \in W$, $\alpha^{(2)} \in C$, and $\check{\alpha} = \alpha^{(1)}\alpha^{(2)}$. Hence $\alpha = \hat{\alpha}\alpha^{(1)}\alpha^{(2)} \in WC$. It remains to show that W is D-invariant. For all $\gamma \in V$, $\beta \in D$ we have $\gamma^\beta \in V$. Therefore, putting $(\alpha_{ij}) := \mathfrak{p}(\beta^{-1})$, we have

$$\gamma^{\beta\varphi} = \sum_{j,k=1}^{2} \alpha_{2j}\gamma_{jk}\beta_{k2} + J(E) = \alpha_{22}\gamma_{22}\beta_{22} + J(E),$$

by Lemma 3(ii). Now if $\gamma \in W$, then there exists an element $\zeta \in J(E)$ such that $\alpha_{22} = \mathrm{id}_{G_2} + \zeta$. Therefore, Lemma 3(iii) shows that it suffices to prove

(14) $\alpha_{22}\zeta\beta_{22} \in J(E)$ for all $\zeta \in J(E)$.

Let $\delta \in E$. Then for all $\omega \in \Omega$ we have, by Lemma 3(i),

$$\omega|_{G_2}\beta_{22}\delta\alpha_{22} = \beta_{22}\omega^{\beta}\delta\alpha_{22} = \beta_{22}\delta\alpha_{22}\omega^{\beta\alpha} = \beta_{22}\delta\alpha_{22}\omega.$$

Hence $\beta_{22}\delta\alpha_{22} \in E$, and therefore $(\alpha_{22}\zeta\beta_{22}\delta)^{n+1} = \alpha_{22}(\zeta\beta_{22}\delta\alpha_{22})^{n}\zeta\beta_{22}\delta = 0$ if n is large. This implies (14), completing the proof of Theorem 2.

As a group may be viewed as a skew ring with trivial multiplication, we obtain the following

Corollary 1. Let G be a group, A a finite abelian normal subgroup of G and $H := C_G(A)$. Put $C := C_{Aut(G)}(G/A)$, $D := N_{Aut(G)}(A)$, $V := C_D(G/H)$. Then V contains a D-invariant subgroup W such that $W \cap C$ is nilpotent and $WC = V$.

Theorem I in our Introduction is the special case $A = Z(G)$ of this corollary. For (not necessarily associative) rings, Theorem 2 yields, e.g.,

Corollary 2. Let G be an algebra over a commutative ring R, A a zero R-ideal of G satisfying both chain conditions for R-ideals, $H := Ann_G(A)$. Put $C := C_{Aut_R(G)}(G/A)$, $D := N_{Aut_R(G)}(A)$, $V := C_D(G/H)$. Then V contains a D-invariant subgroup W such that $W \cap C$ is nilpotent and $WC = V$.

If $A = Ann(G)$, this result becomes Theorem II of our Introduction.

References

1. H. Fitting, Die Gruppe der zentralen Automorphismen einer Gruppe mit Hauptreihe, Math. Ann. 114 (1937), 355-372.
2. H. Heineken and H. Liebeck, The occurence of finite groups in the automorphism group of nilpotent groups of class 2, Arch. Math. 25 (1974), 8-16.
3. H. Laue, On group automorphisms which centralize the factor group by an abelian normal subgroup, J. Algebra 96 (1985), 532-547.
4. H. Laue, Direct decompositions of ideals and normal subgroups, to appear in Adv. in Math.
5. O. Taussky, Rings with non-commutative addition, Bull. Calcutta Math. Soc. 28 (1936), 245-246.

ALGEBRAICALLY CLOSED GROUPS IN LOCALLY FINITE GROUP CLASSES

Felix Leinen and Richard E. Phillips

Fachbereich 17 - Mathematik
Johannes-Gutenberg-Universität
Saarstr. 21
6500 Mainz
WEST GERMANY

Department of Mathematics
Wells Hall
Michigan State University
East Lansing, Michigan 48824
U.S.A.

I. Introduction.

1.1. The concept of existentially closed groups appears to be playing an increasingly important role in the theory of infinite groups. Recall that, for a class $\underline{\underline{X}}$ of groups, a group $G \in \underline{\underline{X}}$ is existentially closed (e.c.) in $\underline{\underline{X}}$, if

(1.1.1) every system S of finitely many equations and inequalities with coefficients in G, which has a solution in some $H \in \underline{\underline{X}}$ with $G \leq H$,

also has a solution in G. A system S of finitely many equations and inequalities satisfying (1.1.1) is said to be $\underline{\underline{X}}$-consistent over G.

If $\underline{\underline{X}}$ is inductive (closed under unions of ascending chains), then every $X \in \underline{\underline{X}}$ is contained in an e.c. $\underline{\underline{X}}$-group G; and if $\underline{\underline{X}}$ is also subgroup closed, then G can be so chosen that $|G| \leq \max\{\aleph_0, |X|\}$ [6, Proposition I.1.3]. The e.c. $\underline{\underline{X}}$-groups (for various classes $\underline{\underline{X}}$) tend to be universal with respect to $\underline{\underline{X}}$ and frequently display properties that are very difficult to duplicate by other methods of construction (see, for example [4], [9], [11], [13]).

A close relative of the e.c. $\underline{\underline{X}}$-groups are the algebraically closed $\underline{\underline{X}}$-groups. A group $G \in \underline{\underline{X}}$ is algebraically closed (a.c.) in $\underline{\underline{X}}$, if every system S of finitely many equations, which is $\underline{\underline{X}}$-consistent over G, has a solution in G. Obviously, every e.c. $\underline{\underline{X}}$-group is a.c. in $\underline{\underline{X}}$. Note, however, that the trivial group is a.c. in the class of all groups, and so a.c. $\underline{\underline{X}}$-groups need not be e.c. in $\underline{\underline{X}}$. Less trivial examples of a.c. $\underline{\underline{X}}$-groups that are not e.c. in $\underline{\underline{X}}$ can be found in the class $\underline{\underline{A}} \cap \underline{\underline{P}}$ of all periodic Abelian groups. Any divisible $(\underline{\underline{A}} \cap \underline{\underline{P}})$-group is a.c. in $\underline{\underline{A}} \cap \underline{\underline{P}}$, while there is a unique countable e.c. $(\underline{\underline{A}} \cap \underline{\underline{P}})$-group; namely the divisible $(\underline{\underline{A}} \cap \underline{\underline{P}})$-group with countably infinite p-rank for every prime p.

Since the defining conditions for a.c. $\underline{\underline{X}}$-groups are weaker than those for e.c. $\underline{\underline{X}}$-groups, one expects the class $\underline{\underline{X}}$ to contain more a.c. groups than e.c. groups. Further, one anticipates tighter structural properties in e.c. $\underline{\underline{X}}$-groups. On the other hand, a.c. $\underline{\underline{X}}$-groups should be easier to recognize. In view of these remarks, the result of B.H.Neumann [14], which asserts that every non-trivial a.c. group (in the class of all groups) is an e.c. group, comes as a bit of a surprise.

In this paper we follow the method of B.H.Neumann and delineate the differences between a.c. and e.c. structures in various classes of groups. Of course, considerable information must be available regarding the a.c. groups in the relevant classes, and the availability of such informations has dictated the classes we study. We define and briefly discuss these classes in the following section.

<u>1.2. The Classes of Groups and Statements of Results</u>. The relevant symbols for the classes we study, together with references regarding the e.c. groups in these classes, are given in

(1.2.1) (a) $L\underline{\underline{F}}$ = (locally finite groups) ; [2] , [7, § 6] , [12] .

(b) For any prime p,
$L\underline{\underline{F}}_p$ = (locally finite p-groups) ; [10] , [13] .

(c) For any set π of primes,
$L(\underline{\underline{F}}_\pi \cap \underline{\underline{S}})$ = (periodic locally soluble π-groups); [8],[9].

(d) For any periodic Abelian group A,
$(L\underline{\underline{F}} ; A \leq \zeta)$ = ($L\underline{\underline{F}}$-groups whose centre contains A); [4],[16].

(e) For any prime p and any Abelian p-group A,
$(L\underline{\underline{F}}_p ; A \leq \zeta)$ = ($L\underline{\underline{F}}_p$-groups whose centre contains A) ; [11] .

(f) For any set π of primes and any Abelian π-group A,
$(L(\underline{\underline{F}}_\pi \cap \underline{\underline{S}}) ; A \leq \zeta)$ = ($L(\underline{\underline{F}}_\pi \cap \underline{\underline{S}})$-groups whose centre contains A).

(g) For any set π of primes and any $p \in \pi$,
$(L(\underline{\underline{F}}_\pi \cap \underline{\underline{S}}) ; C_p)$ = ($L(\underline{\underline{F}}_\pi \cap \underline{\underline{S}})$-groups that have a minimal normal subgroup containing C_p).

It is essential that in the classes defined in (d),(e),(f) and (g) the subgroup A (respectively C_p) remains fixed for different members of the corresponding class. For example, if $G, H \in (L\underline{\underline{F}} ; A \leq \zeta)$ and $G \leq H$, then $A \leq \zeta(G)$ equals $A \leq \zeta(H)$. Therefore, each of the above classes is inductive and closed with respect to subgroups (containing A (respectively C_p)). Finally note that, for any prime p,

(1.2.2) $L(\underline{\underline{F}}_{\{p\}} \cap \underline{\underline{S}}) = L\underline{\underline{F}}_p$ and $(L(\underline{\underline{F}}_{\{p\}} \cap \underline{\underline{S}}) ; C_p) = (L\underline{\underline{F}}_p ; C_p \leq \zeta)$,

since minimal normal subgroups of $L\underline{\underline{F}}_p$-groups are central [7 ; 1.B.8].

We now move on to the statements of our results.

Theorem A. (a) The non-trivial a.c. $L\underline{\underline{F}}$-groups are precisely the e.c. $L\underline{\underline{F}}$-groups.

(b) For any non-trivial periodic Abelian group A, the a.c. $(L\underline{\underline{F}}\,;\,A\leq\zeta)$-groups are precisely the e.c. $(L\underline{\underline{F}}\,;\,A\leq\zeta)$-groups.

Theorem B. (a) A non-trivial group is a.c. in $L\underline{\underline{F}}_p$ if and only if it is either e.c. in $L\underline{\underline{F}}_p$ or e.c. in $(L\underline{\underline{F}}_p\,;\,C_p\leq\zeta)$.

(b) For any non-trivial Abelian p-group A, the a.c. $(L\underline{\underline{F}}_p\,;\,A\leq\zeta)$-groups are precisely the e.c. $(L\underline{\underline{F}}_p\,;\,A\leq\zeta)$-groups.

Since there exists a unique countable e.c. group in each of the classes $L\underline{\underline{F}}_p$ [13, Satz 2] and $(L\underline{\underline{F}}_p\,;\,C_p\leq\zeta)$ [11, Corollary 1], we conclude easily from Theorem B that

(1.2.3) there are exactly two isomorphism types of non-trivial countable a.c. $L\underline{\underline{F}}_p$-groups.

The analogue of Theorem B for locally soluble π-groups is given in

Theorem C. (a) If G is either a.c. in $(L(\underline{\underline{F}}_\pi\cap\underline{\underline{S}})\,;\,C_p)$ or e.c. in $L(\underline{\underline{F}}_\pi\cap\underline{\underline{S}})$, then G is a.c. in $L(\underline{\underline{F}}_\pi\cap\underline{\underline{S}})$.

(b) If the a.c. $L(\underline{\underline{F}}_\pi\cap\underline{\underline{S}})$-group G has a (non-trivial) minimal normal p-subgroup, then G is a.c. in $(L(\underline{\underline{F}}_\pi\cap\underline{\underline{S}})\,;\,C_p)$.

(c) If the non-trivial countable a.c. $L(\underline{\underline{F}}_\pi\cap\underline{\underline{S}})$-group G has no minimal normal subgroups, then G is e.c. in $L(\underline{\underline{F}}_\pi\cap\underline{\underline{S}})$.

(d) The a.c. $(L(\underline{\underline{F}}_\pi\cap\underline{\underline{S}})\,;\,C_p)$-groups are precisely the e.c. $(L(\underline{\underline{F}}_\pi\cap\underline{\underline{S}})\,;\,C_p)$-groups.

(e) For any non-trivial Abelian π-group A, the a.c. $(L(\underline{\underline{F}}_\pi\cap\underline{\underline{S}});A\leq\zeta)$-groups are precisely the e.c. $(L(\underline{\underline{F}}_\pi\cap\underline{\underline{S}})\,;\,A\leq\zeta)$-groups.

Since minimal normal subgroups of $L(\underline{\underline{F}}_\pi\cap\underline{\underline{S}})$-groups are elementary Abelian p-groups for some $p\in\pi$ [7, 1.B.6], and since the e.c. $L(\underline{\underline{F}}_\pi\cap\underline{\underline{S}})$-groups have no minimal normal subgroups [9, Theorem 2.6], Theorem C(a) implies that

(1.2.4) there are at least $|\pi|+1$ isomorphism types of non-trivial countable a.c. $L(\underline{\underline{F}}_\pi\cap\underline{\underline{S}})$-groups.

Further, parts (b) and (c) of Theorem C combine with part (a) to give a complete description of the countable a.c. $L(\underline{\underline{F}}_\pi\cap\underline{\underline{S}})$-groups. Clearly, Theorem C(c) should also be true without the countability assumption. However, we do not know how to prove this. The crucial point is that we have not been able to show that every finite subgroup of an e.c. $L(\underline{\underline{F}}_\pi \cap \underline{\underline{S}})$-group G is contained in a countable subgroup of G which is e.c. in $L(\underline{\underline{F}}_\pi \cap \underline{\underline{S}})$. (This holds for e.c. $L\underline{\underline{F}}_p$-groups [10, Theorem 3.3].)

As a final byproduct of Theorem C(a), we have the embedding result

(1.2.5) If the $L(\underline{\underline{F}}_\pi \cap \underline{\underline{S}})$-group S has a minimal normal p-subgroup K, then S is contained in an a.c. $L(\underline{\underline{F}}_\pi \cap \underline{\underline{S}})$-group G of cardinality $\max\{\aleph_0, |S|\}$, which has a minimal normal subgroup M containing K.

To obtain (1.2.5), choose $1 \neq t \in K$ and identify $\langle t \rangle$ with C_p. The standard existence argument [6, Proposition I.1.3] produces an e.c. $(L(\underline{\underline{F}}_\pi \cap \underline{\underline{S}}) ; C_p)$-group G with $S \leq G$ and $|G| = \max\{\aleph_0, |S|\}$. By Theorem C(a), the group G is a.c. in $L(\underline{\underline{F}}_\pi \cap \underline{\underline{S}})$. And finally, $K = \langle t^S \rangle \leq \langle t^G \rangle = M$.

The a.c. $L(\underline{\underline{F}}_\pi \cap \underline{\underline{S}})$-groups admit a characterization as certain images of e.c. $L(\underline{\underline{F}}_\pi \cap \underline{\underline{S}})$-groups, and this fact is the content of our

Theorem D. (a) For every a.c. $L(\underline{\underline{F}}_\pi \cap \underline{\underline{S}})$-group G with a (non-trivial) minimal normal p-subgroup, there exists an e.c. $L(\underline{\underline{F}}_\pi \cap \underline{\underline{S}})$-group H with $|H| = |G|$ such that $G \cong H/N$ for some proper normal subgroup N of H satisfying $N \nsubseteq \langle h^H \rangle$ for all $h \in H$.

(b) The a.c. $L\underline{\underline{F}}_p$-groups are precisely the factor groups H/N of the e.c. $L\underline{\underline{F}}_p$-groups H by their normal subgroups N satisfying $N \nsubseteq \langle h^H \rangle$ for all $h \in H-1$.

(c) The countable a.c. $L(\underline{\underline{F}}_\pi \cap \underline{\underline{S}})$-groups are precisely the factor groups H/N of the countable e.c. $L(\underline{\underline{F}}_\pi \cap \underline{\underline{S}})$-groups H by their normal subgroups N satisfying $N \nsubseteq \langle h^H \rangle$ for all $h \in H-1$.

Again we conjecture, that the countability assumption in (c) can be removed, but our techniques seem to be not strong enough to prove this.

1.3. Structure of the Paper and some Preliminary Results. Many of our results are of the form "G a.c. in $\underline{\underline{X}}$ implies G e.c. in $\underline{\underline{X}}$" (for various choices of $\underline{\underline{X}}$). We outline a general method of obtaining such a result. Let

(1.3.1) $S = E \cup I$

be a system of equations E and inequalities I which is $\underline{\underline{X}}$-consistent over the $\underline{\underline{X}}$-group G. Our technique will provide a system T of equations which is $\underline{\underline{X}}$-consistent over G and each of whose solutions also yields a solution of S. This is essentially the method of B.H.Neumann [14]. Our variant of this technique will be presented in Section 2.1 (as Lemmata 1, 2 and 3). These results will suffice to establish Theorem A (Section 2.2).

Subsequent to the proof of Theorem A, we will concentrate on Theorem C, since much of Theorem B follows from Theorem C. The proof of Theorem C will require results about the structure of a.c. groups in various periodic locally soluble group classes; it will be necessary to develop these results herein. In many cases, such results will follow from known arguments concerning the e.c. groups in related classes. Basically, such arguments only use the ability to solve equations over a group, and so apply to a.c. structures as well as to e.c. structures. In regard to these points, our work is definitely not self-contained as we will make repeated references to [8], [9] and [11].

We collectively present the structural results about a.c. groups in the following proposition, which will be proved in Section 3.2.

Proposition. *Let* G *be a non-trivial a.c.* $L(\underline{\underline{F}}_\pi \cap \underline{\underline{S}})$*-group*.

(a) G *is verbally complete*.

(b) *If* M/N *is a chief factor in* G, *then for every non-trivial word* $w(x_1,\dots,x_n)$ *and for every* $h \in N$ *there exist* $g_1,\dots,g_n \in M$ *such that* $h = w(g_1,\dots,g_n)$. *In particular*, $N = M'$. *Also*, $M = \langle g^G \rangle$ *for every* $g \in M-N$.

(c) G *has exactly one chief series. Thus, the normal subgroups of* G *form a chain. The order type of the unique chief series is a dense order with no maximal element*.

(d) *If* M/N *is a p-chief factor in* G, *then* M-N *contains an element of order p*.

(e) *If* $|\pi| \neq 1$, *then every chief factor of* G *is infinite (and hence not central)*.

(f) *For every* $p \in \pi$ *and for any two chief factors* M_1/N_1 *and* M_2/N_2 *in* G *with* $M_2 \leqq N_1$, *there exists a p-chief factor* M/N *in* G *such that* $M_2 \leqq N \triangleleft M \leqq N_1$.

The definitions of *chief series*, *chief factors* and *order type* can be found in [9, § 2]. All parts of the Proposition have analogues for a.c. $(L(\underline{\underline{F}}_\pi \cap \underline{\underline{S}})\,;\, A \leqq \zeta)$-groups and chief factors M/N with $A \leqq N$. Proofs of such results rely on techniques of [11]. Also, results and proofs for $L(\underline{\underline{F}}_\pi \cap \underline{\underline{S}})$-groups can be carried out in the slightly more general setting of an abstract class of locally-$\underline{\underline{X}}$-groups, where $\underline{\underline{X}}$ is a class of $L\underline{\underline{F}}$-groups satisfying certain closure operations (as in [9, § 1]).

The proof of Thoorom C will be given in Section 4.1, the remaining parts of Theorem B will be cleaned up in Section 4.2, while Theorem D is proved in Section 4.4.

II. Fundamental Lemmata and the Proof of Theorem A.

2.1. Lemmata. We assume troughout that a system of finitely many equations and inequalities with coefficients in G is of the form

(2.1.1) $S = E \cup I$, where

$$E = \{ w_i(x_1,\dots,x_n;g_1,\dots,g_m) = 1 \mid 1 \leqq i \leqq s \} \quad \text{and}$$

$$I = \{ v_j(x_1,\dots,x_n;g_1,\dots,g_m) \neq 1 \mid 1 \leqq j \leqq t \} .$$

Here, $x_1,\dots,x_n$ are variables and $g_1,\dots,g_m$ are elements of G. We will also use the vector form $\bar{x} = (x_1,\dots,x_n)$ and $\bar{g} = (g_1,\dots,g_m)$. Then

$$E = \{ w_i(\bar{x},\bar{g}) = 1 \mid 1 \leqq i \leqq s \} \quad \text{and} \quad I = \{ v_j(\bar{x},\bar{g}) \neq 1 \mid 1 \leqq j \leqq t \} .$$

A solution of S in some group $H \geqq G$ will be denoted by $\bar{h} = (h_1,\dots,h_n)$.

Lemma 1. Let $G \leqq H$ be groups and suppose that H contains a solution $\bar{h}$ of the system S of (2.1.1). Assume also that

(2.1.2) there exist elements $1 \neq d_j \in \langle v_j(\bar{h},\bar{g})^H \rangle \cap G$ for $1 \leqq j \leqq t$.

Then there are positive integers n_j, $1 \leqq j \leqq t$, and $\varepsilon_{jr} \in \{-1,1\}$, $1 \leqq r \leqq n_j$, such that the system of equations

(2.1.3) $T = E \cup E'$, where

$$E' = \{ d_j^{-1} \cdot \prod_{r=1}^{n_j} v_j(\bar{x},\bar{g})^{\varepsilon_{jr} y_{jr}} = 1 \mid 1 \leqq j \leqq t \}$$

with unknowns $\bar{x}$ and $\bar{y} = (y_{jr} \mid 1 \leqq r \leqq n_j , 1 \leqq j \leqq t)$, and with coefficients $\bar{g}$ and $\bar{d} = (d_1,\dots,d_t)$ from G, has a solution in H. Moreover, every solution of T is also a solution of S.

Proof. The hypothesis (2.1.2) implies that T has a solution in H for an appropriate choice of n_j and ε_{jr}. Suppose that $\bar{h}$ (substituted for $\bar{x}$) and $\bar{k}$ (substituted for $\bar{y}$) provide a solution for T. Clearly then, $\bar{h}$ is a solution for E. Further, $d_j \neq 1$ forces $v_j(\bar{h},\bar{g}) \neq 1$ for $1 \leqq j \leqq t$. Thus, $\bar{h}$ is a solution of S, completing the proof of Lemma 1.

The way in which Lemma 1 will be applied can be gleaned from

Lemma 2. Let G be an a.c. $\underline{\underline{X}}$-group. Suppose that, for every $\underline{\underline{X}}$-consistent system S as in (2.1.1) with coefficients $\bar{g}$ in G, there exists an $\underline{\underline{X}}$-group $H \geqq G$ satisfying

(2.1.4) there is a solution $\bar{h}$ of S in H such that

$$\langle v_j(\bar{h},\bar{g})^H\rangle \cap G \neq 1 \quad \underline{\text{for}} \quad 1 \leq j \leq t .$$

Then G is an e.c. $\underline{\underline{X}}$-group.

Proof. By Lemma 1, the system T (as in (2.1.3)) has a solution in $H \in \underline{\underline{X}}$. Since G is an a.c. $\underline{\underline{X}}$-group, T does already have a solution in G. This is a solution to S in G, by Lemma 1.

For many of our arguments, the following weaker version of Lemma 2 will suffice.

Lemma 3. Let G be an a.c. $\underline{\underline{X}}$-group. Suppose that, for every $\underline{\underline{X}}$-consistent system S of finitely many equations and inequalities over G, there exists an $\underline{\underline{X}}$-group $H \geq G$ which contains a solution for S and satisfies

(2.1.5) $\quad \langle z^H\rangle \cap G \neq 1 \quad$ for every $\quad z \in H-1$.

Then G is an e.c. $\underline{\underline{X}}$-group.

2.2. Proof of Theorem A. Put $\underline{\underline{X}} = (L\underline{\underline{F}}\,;\, A \leq \zeta)$. Let G be a.c. in $\underline{\underline{X}}$. In the case when $A = 1$, let $G \neq 1$. Suppose that an $\underline{\underline{X}}$-consistent system of finitely many equations and inequalities over G has a solution in some $\underline{\underline{X}}$-group $H \geq G$. Since H is contained in an e.c. $\underline{\underline{X}}$-group, we may assume without loss that H is e.c. in $\underline{\underline{X}}$.

Let $z \in H-1$. If $z \in A$, then $z \in G$. Suppose now that $z \in H-A$. By [4, Theorem 1(a)], H/A is simple. Thus, H is the central product of $\langle z^H\rangle$ and A. But H is also perfect [16, (3.1.1)]. Hence, $\langle z^H\rangle = H$. This shows that $\langle z^H\rangle \cap G \neq 1$ for every $z \in H-1$. Therefore, Lemma 3 applies.

III. Some Remarks about Amalgams and the Proof of the Proposition.

3.1. Amalgams. The proof of the Proposition as well as several other arguments in this paper require a construction that produces completions of certain group amalgams. This construction is essentially that of [9, Theorem 2.1]. It will be used tacitly in many situations (for example, when we quote certain arguments from [9] or [11]), but there will be other occasions when it will be necessary to use this construction in a more direct way. At any rate, our presentation will be somewhat more complete with the inclusion of a statement of this result.

(3.1.1) Let G,H,U be $L(\underline{\underline{F}}_\pi \cap \underline{\underline{S}})$-groups with U finite. Let

$$\underline{\underline{A}}: \quad G \xleftarrow{\alpha} U \xrightarrow{\beta} H$$

be an <u>amalgam</u> (which means that α and β are embeddings). Suppose further that there exist normal series Σ_G in G and Σ_H in H such that

(i) $(\Sigma_G \cap U\alpha)\alpha^{-1} = (\Sigma_H \cap U\beta)\beta^{-1}$, and

(ii) $K/L \leqq (U\beta)L/L \cap \zeta(H/L)$ for all $(K,L) \in \Sigma_H$ with $K \cap U\beta \neq L \cap U\beta$.

Then there exist $J \in L(\underline{\underline{F}}_\pi \cap \underline{\underline{S}})$ and embeddings $\gamma: G \to J$ and $\delta: H \to J$ such that the diagram

$$\begin{array}{ccccc} G & \xleftarrow{\alpha} & U & \xrightarrow{\beta} & H \\ & \gamma \searrow & & \swarrow \delta & \\ & & J & & \end{array}$$

commutes.

The group J of (3.1.1) (called a <u>completion</u> of $\underline{\underline{A}}$) is a permutational product of the amalgam $\underline{\underline{A}}$ with respect to a certain choice of transversals of $U\alpha$ in G and of $U\beta$ in H. This same transversal choice is used in [11, Proposition 1] in a more restrictive setting. With this observation in mind, [11, Proposition 1] can be easily modified as

(3.1.2) In addition to the hypotheses of (3.1.1) suppose also that

(iii) $A \leqq \zeta(G)$,

(iv) $(U\alpha \cap A)\alpha^{-1}\beta \leqq \zeta(H)$, and

(v) Σ_G refines the normal series $\{(G,A),(A,1)\}$.

Then there exists a completion $J \in L(\underline{\underline{F}}_\pi \cap \underline{\underline{S}})$ of $\underline{\underline{A}}$ as in (3.1.1) and with the additional property that $A\gamma \leqq \zeta(J)$.

<u>3.2. Proof of the Proposition</u>. The proof of <u>parts (a) and (b)</u> can be carried out exactly as in [8, Theorem 2.1] and [9, Theorem 2.3] when one notices that only systems of equations are used in the proofs of these theorems.

<u>Part (c)</u>. It is an immediate consequence of (b) that G has a unique chief series and that the normal subgroups of G are totally ordered (see proof of [8, Theorem 4.7]). If M_1/N_1 and M_2/N_2 were two chief factors in G with $M_2 = N_1$, then M_1/N_2 would be metabelian, contrary to the fact that $N_1 = M_1''$ (by (b)). Hence the chief factors of G lie densely. Furthermore, the verbal completeness of G (part (a)) ensures that G has no maximal normal subgroup.

<u>Part (d)</u>. Let $g \in M-N$, where M/N is a p-chief factor in G. We argue

as in [9, Theorem 2.5(a)] (with $V = \{[x_1, x_2]\}$) to find an $L(\underline{\underline{F}}_\pi \cap \underline{\underline{S}})$-group $H \geqq G$ in which the equations

$$g = x_1 \cdot [x_2, x_3] \quad \text{and} \quad x_1^p = 1$$

with coefficient $g \in G$ have a solution $\bar{h} = (h_1, h_2, h_3)$ with

$$h_2, h_3 \in \langle g^H \rangle \ .$$

As in our Lemma 1, the inclusions $h_2, h_3 \in \langle g^H \rangle$ may be viewed as solutions of two additional equations with coefficient $g \in G$. Since G is a.c. in $L(\underline{\underline{F}}_\pi \cap \underline{\underline{S}})$, there exist $g_1, g_2, g_3 \in G$ satisfying

$$g = g_1 \cdot [g_2, g_3] \ , \quad g_1^p = 1 \quad \text{and} \quad g_2, g_3 \in \langle g^G \rangle = M \ .$$

Since $[g_2, g_3] \in N$, we see that $g_1 \in gN \subseteq M-N$ is the desired element of order p.

Part (e). Let $|\pi| > 1$, and let M/N be a chief factor in G. Assume that M/N is finite. Then $C_G(M/N)$ is a normal subgroup of finite index in G. Since G has no proper maximal normal subgroup (part (b)), we conclude that $G = C_G(M/N)$. Thus, $M/N \leqq \zeta(G/N)$ and $M/N \cong C_p$ for some $p \in \pi$. Let S be any finite soluble π-group with a minimal normal p-subgroup $K \not\leqq \zeta(S)$. From (d), there is an element g of order p in $M-N$. Let $1 \neq x \in K$, and denote by $\mu : \langle g \rangle \longrightarrow \langle x \rangle$ the isomorphism with $g\mu = x$. Since $M/N \leqq \zeta(G/N)$, we can apply the amalgamation result (3.1.1) to the amalgam

$$\underline{\underline{A}}: \quad G \xleftarrow{\ id\ } \langle g \rangle \xrightarrow{\ \mu\ } S$$

to produce an $L(\underline{\underline{F}}_\pi \cap \underline{\underline{S}})$-group $P \geqq G$ and an embedding $\rho : S \longrightarrow P$ such that $x\rho = g$. Because of $K \not\leqq \zeta(S)$ we have $[g, s\rho] \neq 1$ for some $s \in S$. And since K is a minimal normal subgroup of S, we obtain that

$$g \in \langle [g, s\rho]^{S\rho} \rangle \leqq \langle [g, s\rho]^P \rangle \ .$$

Since G is a.c. in $L(\underline{\underline{F}}_\pi \cap \underline{\underline{S}})$, there already exists $h \in G$ with

$$g \in \langle [g, h]^G \rangle \ .$$

But now the centrality of M/N implies that $g \in N$, in contradiction to the choice of g. Therefore, M/N cannot be finite.

Part (f). Let M_1/N_1 and M_2/N_2 be chief factors of G with $M_2 \leqq N_1$. Since the chief factors of G lie densely (part (c)), we can find chief factors M_1^*/N_1^* and M_2^*/N_2^* in G such that

$$N_2 \lhd M_2 < N_2^* \lhd M_2^* < N_1^* \lhd M_1^* < N_1 \lhd M_1 \ .$$

From (d) there exist elements $g_i \in M_i^* - N_i^*$ of prime order. And the argument of [9, Lemma 2.2] shows that we may also assume that $[g_1, g_2] = 1$.

Now, as in the proof of [9, Theorem 2.6(a)], we can find an $L(\underline{\underline{F}}_\pi \cap \underline{\underline{S}})$-group $P \geqq G$ in which the equation $x^p = 1$ has a solution $h \in P$ satisfying

$$g_2 \in \langle h^P\rangle \quad \text{and} \quad h \in \langle g_1^P\rangle \quad .$$

Since G is a.c. in $L(\underline{F}_\pi \cap \underline{S})$, there exists $g \in G$ such that

$$g^p = 1 \quad , \quad g_2 \in \langle g^G\rangle \quad \text{and} \quad g \in \langle g_1^G\rangle \quad .$$

From $g_2 \in \langle g^G\rangle$ we deduce that $g \neq 1$, and so $o(g) = p$. Now, the (unique) chief factor M/N in G with $g \in M-N$ satisfies $N_1^* \leqq N \triangleleft M \leqq M_1^*$, as desired.

<u>3.3. Preliminary Results about</u> $(L(\underline{F}_\pi \cap \underline{S})\ ;\ C_p)$<u>-Groups and</u> $(L(\underline{F}_\pi \cap \underline{S})\ ;\ A \leqq \zeta)$<u>-Groups</u>. As in the proof of Theorem A the proof of Theorem C will use the following lemmata.

<u>Lemma 4</u>. <u>If</u> G <u>is a.c. in</u> $(L(\underline{F}_\pi \cap \underline{S})\ ;\ A \leqq \zeta)$, <u>then</u> $A \leqq \langle g^G\rangle$ <u>for every</u> $g \in G-A$.

<u>Proof</u>. We may assume that $A \neq 1$. Let $g \in G-A$ and $a \in A$. Put $U = \langle g,a\rangle$ and $A_o = U \cap A$. Let

$$\sigma : U \longrightarrow W = A_o \operatorname{Wr} U/A_o$$

be a Krasner-Kaloujnine embedding (see [11, § 3.3]). Identify A_o with the diagonal subgroup of $P = A_o \operatorname{Wr} C$ where C denotes the cyclic group of order $|A_o|$. Then P is a finite soluble π-group with $A_o \leqq \zeta(P) \cap P'$ [15, Theorem 4.1]. Regard W as a subgroup of $H = P \operatorname{Wr} U/A_o$. Then

(3.3.1) $\quad A_o\sigma \leqq \zeta(H) \cap \langle g\sigma^H\rangle \qquad$ [11, (3.2.3)] .

We define normal series $\Sigma_G = \{(G,A),(A,1)\}$ in G and $\Sigma_H = \{(H,\Omega),(\Omega,A_o\sigma),(A_o\sigma,1)\}$ in H ; here, Ω denotes the base group of H. Then $\Sigma_G \cap U = \{(U,A_o),(A_o,1)\}$ and $\Sigma_H \cap U\sigma = \{(U\sigma,A_o\sigma),(A_o\sigma,1)\}$. Further, $A_o\sigma \leqq \zeta(H)$, and H/Ω is Abelian. Thus, the amalgam

$$\underline{A}: \quad G \xleftarrow{\ id\ } U \xrightarrow{\ \sigma\ } H$$

satisfies the conditions of (3.1.2). So from (3.1.2) we obtain an $(L(\underline{F}_\pi \cap \underline{S})\ ;\ A \leqq \zeta)$-group $J \geqq G$ and an embedding $\rho : H \longrightarrow J$ with $\sigma\rho = id(U)$. From (3.3.1) we have $U = U\sigma\rho \leqq \langle g\sigma\rho^{H\rho}\rangle = \langle g^{H\rho}\rangle$. In particular, $a \in \langle g^J\rangle$. But G is a.c. in $(L(\underline{F}_\pi \cap \underline{S})\ ;\ A \leqq \zeta)$, and thus $a \in \langle g^G\rangle$.

<u>Lemma 5</u>. <u>If</u> G <u>is a.c. in</u> $(L(\underline{F}_\pi \cap \underline{S})\ ;\ C_p)$ <u>with minimal normal subgroup</u> $M \geqq C_p$, <u>then</u> $M \leqq \langle g^G\rangle$ <u>for all</u> $g \in G-1$. <u>In particular</u>, M <u>is the unique minimal normal subgroup of</u> G.

<u>Proof</u>. Let $C_p = \langle c\rangle$ and $g \in G-1$. If $\langle g^G\rangle \cap M \neq 1$, then $M \leqq \langle g^G\rangle$

by the minimality of M, as desired. Thus we may assume that

(3.3.2) $\langle g^G\rangle \cap M = 1$, and therefore $[g,c] = 1$.

Put $U = \langle c\rangle\times\langle g\rangle$, and choose P to be a finite p-group with $C_p \cong \langle y\rangle \leq \zeta(P)\cap P'$. Let $H = P\,\mathrm{Wr}\,\langle x\rangle$ where $\langle x\rangle \cong \langle g\rangle$, and define an embedding $\sigma: U \longrightarrow H$ by

$g\sigma = x$ and $c\sigma$ = diagonal element with value y.

It follows from [11, (3.2.3)] that

(3.3.3) $c\sigma \in \langle g\sigma^H\rangle$.

Define normal series $\Sigma_G = \{(G,M),(M,1)\}$ in G and $\Sigma_H = \{(H,\Omega),(\Omega,\langle c\sigma\rangle),(\langle c\sigma\rangle,1)\}$ in H, where Ω denotes the base group of H. Then $(\Sigma_G \cap U)\sigma = \{(U\sigma,\langle c\sigma\rangle),(\langle c\sigma\rangle,1)\} = \Sigma_H \cap U\sigma$, and $\langle c\sigma\rangle \leq \zeta(H)$, while H/Ω is Abelian. Thus, the amalgam

$$\underline{\underline{A}}: \quad G \xleftarrow{\ id\ } U \xrightarrow{\ \sigma\ } H$$

satisfies the hypotheses of (3.1.1), and we obtain an $L(\underline{\underline{F}}_\pi \cap \underline{\underline{S}})$-group $J \geq G$ and an embedding $\tau: H \longrightarrow J$ such that $\sigma\tau = id(U)$. As noted in Section 3.1, the group J is a permutational product of $\underline{\underline{A}}$ with respect to certain transversals S of U in G and T of $U\sigma$ in H. Regard the amalgam

$$\overline{\underline{\underline{A}}}: \quad \overline{G} = G/M \xleftarrow{\ \overline{id}\ } U/U\cap M = \overline{U} = U/\langle c\rangle \xrightarrow{\ \overline{\sigma}\ } H/\langle c\sigma\rangle = \overline{H}$$

where $\overline{id}$ and $\overline{\sigma}$ are the embeddings induced by $id: U \longrightarrow G$ and $\sigma: U \longrightarrow H$. Let $\overline{J}$ be the permutational product of $\overline{\underline{\underline{A}}}$ with respect to the transversals $\overline{S} = SM/M$ and $\overline{T} = T\langle c\sigma\rangle/\langle c\sigma\rangle$. A theorem of R.J.Gregorac [1, Theorem 5.2] asserts that there is an embedding

$$\phi: J \longrightarrow W = M\,\mathrm{Wr}_\Lambda \overline{J}$$

where the action of the top group $\overline{J}$ on the $|\Lambda|$ copies of M in the base group Θ of W is induced by the action of $\overline{J}$ on $\Lambda = \overline{S}\times\overline{T}\times\overline{U}$. From the proof of Gregorac's theorem it can be seen [1, bottom of p.123] that elements from G or $H\tau$ are mapped via ϕ into Θ if and only if they lie in M or $\langle c\sigma\rangle$. Therefore, the normal subgroup $Q = (\Theta \cap J\phi)\phi^{-1}$ of J satisfies

(3.3.4) $Q \cap G = M$ and $Q \cap H\tau = \langle c\sigma\tau\rangle = \langle c\rangle$.

Let L be a J-invariant subgroup of Q maximal with regard to $c \notin L$. From (3.3.4) and the minimality of M (respectively $\langle c\rangle$) it follows that

(3.3.5) $L \cap G = 1$ and $L \cap H\tau = 1$.

Now, if $K \trianglelefteq J$ with $L \lneqq K \leq Q$, then $\langle c^J\rangle \leq K$. Thus $\langle c^J\rangle L/L$ is a minimal normal subgroup of $J^* = J/L$. Identifying $\langle c\rangle L/L$ with C_p , it follows that $J^* \in (L(\underline{\underline{F}}_\pi \cap \underline{\underline{S}})\,;\,C_p)$. Because of (3.3.5) we can identify G and $H\tau$ canonically with GL/L and $(H\tau)L/L$. Then $G \leq J^*$, and

$c \in \langle g^{J^*}\rangle$ by (3.3.3). Since G is a.c. in $(L(\underline{\underline{F}}_\pi \cap \underline{\underline{S}}) ; C_p)$, we already have $c \in \langle g^G\rangle$, contrary to (3.3.2).

IV. Proofs of Theorems B, C and D.

We now have the necessary machinery to proceed to the proofs of the remaining theorems. As noted earlier, a substantial portion of Theorem B follows from Theorem C. For this reason we begin with

4.1. Proof of Theorem C. Part (a). Let G be a.c. in $(L(\underline{\underline{F}}_\pi \cap \underline{\underline{S}}) ; C_p)$ with minimal normal subgroup $M \geqq C_p = \langle c\rangle$. Let S be a system of finitely many equations over G that has a solution in some $L(\underline{\underline{F}}_\pi \cap \underline{\underline{S}})$-group $H \geq G$. Let K/L be a chief factor in H with $c \in K-L$. Then $M \cap L = 1$, and Lemma 5 yields $G \cap L = 1$. Thus, G embeds into H/L, and H/L is an $(L(\underline{\underline{F}}_\pi \cap \underline{\underline{S}}) ; C_p)$-group with minimal normal subgroup K/L containing C_p. Since S consists of equations only, it has a solution in H/L. But G is a.c. in $L(\underline{\underline{F}}_\pi \cap \underline{\underline{S}}) ; C_p)$. Hence S does also have a solution in G. This shows that G is a.c. in $L(\underline{\underline{F}}_\pi \cap \underline{\underline{S}})$.

Part (d). Let G be a.c. in $(L(\underline{\underline{F}}_\pi \cap \underline{\underline{S}}) ; C_p)$. Let S be a system of finitely many equations and inequalities that is $(L(\underline{\underline{F}}_\pi \cap \underline{\underline{S}}) ; C_p)$-consistent over G. Then S has a solution in some e.c. $(L(\underline{\underline{F}}_\pi \cap \underline{\underline{S}}) ; C_p)$-group $H \geq G$. From Lemma 5 we see that $C_p \leqq \langle z^H\rangle$ for every $z \in H-1$. Thus, $G \cap \langle z^H\rangle \geqq C_p \neq 1$ for every $z \in H-1$, and Lemma 3 implies that G is already e.c. in $(L(\underline{\underline{F}}_\pi \cap \underline{\underline{S}}) ; C_p)$.

Part (e). The proof is identical with that of part (d), the essential tool being Lemma 4.

Part (b). Let M be a minimal normal p-subgroup in the a.c. $L(\underline{\underline{F}}_\pi \cap \underline{\underline{S}})$-group G. Fix $1 \neq c \in M$ and identify $\langle c\rangle$ with C_p. Then $G \in (L(\underline{\underline{F}}_\pi \cap \underline{\underline{S}}) ; C_p)$, and it is easily seen that G is a.c. in $(L(\underline{\underline{F}}_\pi \cap \underline{\underline{S}}) ; C_p)$.

Part (c). Let G be a non-trivial countable a.c. $L(\underline{\underline{F}}_\pi \cap \underline{\underline{S}})$-group without minimal normal subgroups. Let $S = E \cup I$ be an $L(\underline{\underline{F}}_\pi \cap \underline{\underline{S}})$-consistent system of finitely many equations and inequalities (as in (2.1.1)) with coefficients $\overline{g} = (g_1, \ldots, g_m)$ in G. Then S has a solution in some countable e.c. $L(\underline{\underline{F}}_\pi \cap \underline{\underline{S}})$-group $H \geq G$. Put

$$G_o = \langle g_1, \ldots, g_m\rangle ,$$

and let Σ be the unique chief series in H (Proposition (c)). Since G has no minimal normal subgroup, $\Sigma \cap G$ has no minimal elements. Thus, Proposition (b) yields the existence of $g_o \in G-1$ such that

(4.1.1) $\quad g_o \in \langle g^G\rangle'$ for every $g \in G_o - 1$.

By [8, Theorem 4.11] the order type of Σ is that of the rational numbers $\mathbb{Q}$. Hence, there exists a mapping $\nu: H-1 \longrightarrow \mathbb{Q}$ defined by $h \in M_{h\nu} - N_{h\nu}$ (where M_q/N_q, $q \in \mathbb{Q}$, are the chief factors of H). Clearly, (4.1.1) implies that $g_o\nu < g\nu$ for all $g \in G_o - 1$. Thus, there exists an irrational number r with $g_o\nu < r < g\nu$ for all $g \in G_o - 1$. By [8, Lemma 4.6] there corresponds a normal subgroup K of H to r which does not occur in any chief factor of H and satisfies

$$M_{g_o\nu} \leqq K \leqq N_{g\nu} \quad \text{for all } g \in G_o - 1 .$$

Clearly, $G_o \cap K = 1$. Further, [9, Theorem 4.1] yields that H splits over K and that G_o is contained in a complement L to K in H. Also, $L \cong H/K$ is an e.c. $L(\underline{\underline{F}}_\pi \cap \underline{\underline{S}})$-group [9, Theorem 4.3]. Thus, since S has coefficients in L and a solution in H, it must also have a solution $\overline{h} = (h_1,\ldots,h_n)$ in L.

Let $z = v_j(\overline{h},\overline{g})$ for some inequality $v_j(\overline{x},\overline{g}) \neq 1$ from I. Then $z \in L$, and since H has a unique chief series, the chief factor M/N with $z \in M-N$ satisfies $K \leqq N$. So, Proposition (b) gives

$$g_o \in K \cap G \leqq N \cap G \leqq \langle z^H\rangle \cap G .$$

In particular, $G \cap \langle z^H\rangle \neq 1$, and an application of Lemma 2 yields that G is e.c. in $L(\underline{\underline{F}}_\pi \cap \underline{\underline{S}})$.

<u>4.2. Proof of Theorem B</u>. Because of parts (a),(b) and (d) of Theorem C we only need to show that any non-trivial a.c. $L\underline{\underline{F}}_p$-group G without minimal normal subgroups is e.c. in $L\underline{\underline{F}}_p$. This is a generalized version of Theorem C(c) for $L\underline{\underline{F}}_p$-groups. Accordingly, the proof parallels that of Theorem C.

Let $S = E \cup I$ be an $L\underline{\underline{F}}_p$-consistent system of finitely many equations and inequalities (as in (2.1.1)) with coefficients $\overline{g} = (g_1,\ldots,g_m)$ in G. Then S has a solution $\overline{h} = (h_1,\ldots,h_n)$ in some e.c. $L\underline{\underline{F}}_p$-group $H \geqq G$. As in the proof of Theorem C(c), there exists $g_o \in G-1$ such that

$$g_o \in \langle g^G\rangle' \quad \text{for every } g \in G_o - 1, \text{ where } G_o = \langle g_1,\ldots,g_m\rangle .$$

In particular, there exists a finite subgroup F of G such that

$$g_o \in \langle g^F\rangle' \quad \text{for every } g \in G_o - 1 .$$

By [10, Theorem 3.3] there is a countable e.c. $L\underline{\underline{F}}_p$-group V with $\langle g_o, G_o, F, h_1,\ldots,h_n\rangle \leqq V \leqq H$. Note that

$$g_o \in \langle g^V\rangle' \quad \text{for every } g \in G_o - 1 ,$$

and that V contains the solution $\overline{h}$ of S. Therefore, we can proceed as in the proof of Theorem C(c), with V in place of H, to find a solution $\overline{u} = (u_1,\dots,u_n)$ of S in V with the additional property that

$$g_0 \in \langle v_j(\overline{u},\overline{g})^V\rangle \cap G \leqq \langle v_j(\overline{u},\overline{g})^H\rangle \cap G$$

for every inequality $v_j(\overline{x},\overline{g}) \neq 1$ from I. Now, Lemma 2 yields that G is e.c. in $L\underline{F}_p$.

4.3. The $\rtimes$-Construction. The proof of Theorem D uses a special construction due to F.Leinen [8, Construction 4.1]. Here, we review relevant terminology and the essential features of the construction.

Let $\theta: G \longrightarrow H$ be a group homomorphism with kernel N. Then $\theta^*: H \longrightarrow G$ is a countermap for θ if

$$[h\cdot(g\theta)]\theta^*\theta = h\theta^*\theta \cdot g\theta \qquad \text{for all } g\in G,\ h\in H.$$

For any θ as above, there always exists a countermap θ^*. And every such countermap θ^* gives rise to a standard embedding

$$\sigma: G \longrightarrow N \operatorname{Wr} H \qquad \text{defined by} \qquad g\sigma = f_g\cdot g\theta$$

where the element f_g in the base group of $N \operatorname{Wr} H$ is given by

$$(h)f_g = h\theta^* \cdot g \cdot [h\cdot(g\theta)]\theta^{*-1} \qquad \text{for all } h\in H.$$

Standard embeddings are a generalization of the notion of Krasner-Kaloujnine embeddings [11, § 3.3] and occur in [5, p.303] and [3, p.487].

Before stating the main result about standard embeddings we need one bit of additional notation. For any group G, we denote by

$$G \rtimes G$$

the split extension of G by G, where G acts on itself via conjugation.

We do now come to the main point.

(4.3.1) [8, Construction 4.1]. Let G be the union of an ascending chain $\{G_n\}_{n\in\mathbb{N}}$ of subgroups, and let $\theta: G \longrightarrow H$ be a homomorphism with kernel N. Assume that

$$\sigma_1: G_1 \longrightarrow (G_1 \cap N) \operatorname{Wr} H$$

is a standard embedding with respect to a countermap $\theta_1^*: H \longrightarrow G_1$ of $\theta_1 = \theta\restriction G_1$. Identify $(G_1 \cap N) \operatorname{Wr} H$ in the natural way with the subgroup $[(G_1\cap N)\rtimes 1] \operatorname{Wr} H$ of $(G\rtimes G) \operatorname{Wr} H$.

Then σ_1 can be extended to an embedding

$$\sigma: G \longrightarrow W = \bigcup_{n\in\mathbb{N}} [(G_n \rtimes G_n) \operatorname{Wr} H] \leqq (G\rtimes G) \operatorname{Wr} H$$

in such a manner that, for every $g\in G$, the projection of $g\sigma$

onto the top group of $(G \rtimes G)\,\mathrm{Wr}\,H$ is $g\theta$.

4.4. Proof of Theorem D. Part (a). Let G be a.c. in $L(\underline{\underline{F}}_\pi \cap \underline{\underline{S}})$ with a minimal normal p-subgroup $M \neq 1$. Beginning with $G_1 = G$ and $N_1 = 1$ we will inductively construct ascending chains $\{G_k\}_{k\in\mathbb{N}}$ and $\{N_k\}_{k\in\mathbb{N}}$ of $L(\underline{\underline{F}}_\pi \cap \underline{\underline{S}})$-groups satisfying

(4.4.1) (i) $|G_k| = |G|$,

(ii) $N_k \trianglelefteq G_k$, and G_k splits over N_k with complement G, and

(iii) every system of finitely many equations and inequalities with coefficients from G_k, which is $L(\underline{\underline{F}}_\pi \cap \underline{\underline{S}})$-consistent over G_{k+1}, has a solution in G_{k+1}.

The remainder of part (a) follows from (4.4.1); to see this, put $H = \bigcup\{G_k \mid k\in\mathbb{N}\}$ and $N = \bigcup\{N_k \mid k\in\mathbb{N}\}$. From (iii) of (4.4.1) we see that H is an e.c. $L(\underline{\underline{F}}_\pi \cap \underline{\underline{S}})$-group, while we can easily deduce from (ii) of (4.4.1) that H splits over N with complement G. Further, (4.4.1)(i) yields $|H| = |G|$, since G is infinite. Note also that $N \neq \langle h^H \rangle$ for all $h \in H$ by Proposition (b) and [9, Theorem 2.6(b)] (since G has a minimal normal subgroup).

The proof of (4.4.1) is by induction on k. For the step $k \to k+1$, let $\{S_\alpha \mid \alpha < |G|\}$ be a list of all systems of finitely many equations and inequalities over G_k. Put $H_1 = G_k$ and $L_1 = N_k$. We will construct ascending chains $\{H_\alpha\}_{\alpha<|G|}$ and $\{L_\alpha\}_{\alpha<|G|}$ of $L(\underline{\underline{F}}_\pi \cap \underline{\underline{S}})$-groups such that

(4.4.2) (i) $|H_\alpha| = |G|$,

(ii) $L_\alpha \trianglelefteq H_\alpha$, and H_α splits over L_α with complement G, and

(iii) if S_α is $L(\underline{\underline{F}}_\pi \cap \underline{\underline{S}})$-consistent over H_α, then S_α has a solution in $H_{\alpha+1}$.

Before proceeding with the proof of (4.4.2), notice that (4.4.2) implies (4.4.1) if we put $G_{k+1} = \bigcup\{H_\alpha \mid \alpha < |G|\}$ and $N_{k+1} = \bigcup\{L_\alpha \mid \alpha < |G|\}$. Thus, verification of (4.4.2) will complete the proof of part (a).

We prove (4.4.2) by transfinite induction on α. If $\lambda < |G|$ is a limit ordinal, then let $H_\lambda = \bigcup\{H_\alpha \mid \alpha<\lambda\}$ and $L_\lambda = \bigcup\{L_\alpha \mid \alpha<\lambda\}$. Now suppose, that H_α and L_α have been found for some $\alpha < |G|$. Choose $H_{\alpha+1} = H_\alpha$ and $L_{\alpha+1} = L_\alpha$, if S_α is not $L(\underline{\underline{F}}_\pi \cap \underline{\underline{S}})$-consistent over H_α. Otherwise, we may assume that S_α has coefficients $\overline{g} = (g_1,\dots,g_m)$ in H_α and a solution $\overline{h} = (h_1,\dots,h_n)$ in some $L(\underline{\underline{F}}_\pi \cap \underline{\underline{S}})$-group $\tilde{H} \geq H_\alpha$. Put

$$U_0 = \langle g_1,\dots,g_m\rangle \quad \text{and} \quad U = \langle h_1,\dots,h_n,U_0\rangle \quad .$$

From parts (b) and (d) of Theorem C, the group G is e.c. in $(L(\underline{\underline{F}}_\pi \cap \underline{\underline{S}})\,;\,C_p)$ with minimal normal subgroup $M \geq C_p = \langle c\rangle$. Let $\tilde{K}/\tilde{L}$ be a chief factor in $\tilde{H}$ with $c \in \tilde{K}-\tilde{L}$. Since M is the unique minimal normal subgroup of G (Lemma 5), we have

$$\tilde{L} \cap G = 1 < M \leqq \tilde{K} \cap G \quad .$$

Thus, G can be identified with the subgroup $G\tilde{L}/\tilde{L}$ of $\tilde{H}/\tilde{L}$. Then $\tilde{H}/\tilde{L}$ is an $(L(\underline{\underline{F}}_\pi \cap \underline{\underline{S}}) ; C_p)$-supergroup of G with minimal normal subgroup $\tilde{K}/\tilde{L}$ containing C_p. Since G is e.c. in $(L(\underline{\underline{F}}_\pi \cap \underline{\underline{S}}) ; C_p)$, we obtain an extension

$$\phi : U\tilde{L}/\tilde{L} \longrightarrow G \quad \text{of} \quad id : U_o\tilde{L}/\tilde{L} \longrightarrow G$$

(the group table of $U\tilde{L}/\tilde{L}$ with $U_o\tilde{L}/\tilde{L}$ as constants can be duplicated in G).

Denote by $\theta : U \longrightarrow H_\alpha$ the composition of the canonical epimorphism $U \longrightarrow U\tilde{L}/\tilde{L}$ and $\phi : U\tilde{L}/\tilde{L} \longrightarrow G \leqq H_\alpha$. Let

$$\sigma_1 : U_o \longrightarrow (U_o \cap \tilde{L}) \text{ Wr } H_\alpha = 1 \text{ Wr } H_\alpha$$

be a standard embedding with respect to any countermap θ_1^* for $\theta_1 = \theta \restriction U_o$. Then $\sigma_1 = id(U_o)$, and (4.3.1) yields an embedding

$$\sigma : U \longrightarrow (U \rtimes U) \text{ Wr } H_\alpha = W$$

that extends $id(U_o)$. Obviously, $W \in L(\underline{\underline{F}}_\pi \cap \underline{\underline{S}})$, and $\bar{h}\sigma = (h_1\sigma, \dots, h_n\sigma)$ is a solution of S_α in W.

Now, $h_i\sigma = f_i \cdot (h_i\theta)$ for $1 \leqq i \leqq n$, where f_i is in the base group of W. Put

$$B = \langle f_i^{H_\alpha} \mid 1 \leqq i \leqq n \rangle \quad ,$$

and let $L_{\alpha+1} = B \cdot L_\alpha$ and $H_{\alpha+1} = L_{\alpha+1} \cdot G$. Clearly,

$$H_{\alpha+1} = L_{\alpha+1} \cdot G = B \cdot L_\alpha \cdot G = B \cdot H_\alpha \quad ;$$

thus $B \trianglelefteq H_{\alpha+1}$ and $L_{\alpha+1} \trianglelefteq H_{\alpha+1}$. Further, since B is contained in the base group of W and since H_α splits over L_α, we see that $H_{\alpha+1}$ splits over $L_{\alpha+1}$ with complement G. Because of $|H_\alpha| = |G| = |B|$ we also have $|H_{\alpha+1}| = |G|$. Finally, $H_{\alpha+1} = B \cdot H_\alpha$ contains the solution $\bar{h}\sigma$ of S_α. This completes the proof of (4.4.2).

Part (b). Let G be an a.c. $L\underline{\underline{F}}_p$-group. If $G = 1$, then $G \cong H/N$ for every e.c. $L\underline{\underline{F}}_p$-group H and $N = H$; observe that this choice of N meets the requirements of (b), since $H \neq \langle h^H \rangle$ for all $h \in H$ by part (c) of the Proposition. We may assume now that $G \neq 1$. Then Theorem B yields that G is either e.c. in $L\underline{\underline{F}}_p$ or e.c. in $(L\underline{\underline{F}}_p ; C_p \leqq \zeta)$. In the first case we choose $H = G$ and $N = 1$. In the second case, we may apply (a) to obtain an e.c. $L\underline{\underline{F}}_p$-group H and $N \trianglelefteq H$ such that $G \cong H/N$ and $N \neq \langle h^H \rangle$ for all $h \in H$.

For the converse, let H be an e.c. $L\underline{\underline{F}}_p$-group and $N \ntrianglelefteq H$ with $N \neq \langle h^H \rangle$ for all $h \in H-1$. If N does not occur in any chief factor of H, then H/N is e.c. in $L\underline{\underline{F}}_p$ by [10, Theorem 4.1(d)] . Otherwise, we conclude from part (b) of the Proposition that there exists $M \trianglelefteq H$ such that M/N is a chief factor in H. Now [10, Theorem 4.2] yields that H/N is e.c. in $(L\underline{\underline{F}}_p ; C_{p^{\leq \zeta}})$, and thus H/N is an a.c. $L\underline{\underline{F}}_p$-group by Theorem B.

Part (c). Let G be a countable a.c. $L(\underline{\underline{F}}_\pi \cap \underline{\underline{S}})$-group. Then we can proceed as in the proof of (b), with Theorem C(c) in place of Theorem B, to obtain a countable e.c. $L(\underline{\underline{F}}_\pi \cap \underline{\underline{S}})$-group H and $N \trianglelefteq H$ such that $G \cong H/N$ and $N \nleq \langle h^H \rangle$ for all $h \in H-1$.

For the converse, let H be a countable e.c. $L(\underline{\underline{F}}_\pi \cap \underline{\underline{S}})$-group and $N \ntrianglelefteq H$ with $N \nleq \langle h^H \rangle$ for all $h \in H-1$. If N does not occur in any chief factor of H, then H/N is e.c. in $L(\underline{\underline{F}}_\pi \cap \underline{\underline{S}})$ by [9, Theorem 4.3] . So we may assume again that there exists $M \trianglelefteq H$ such that M/N is a chief factor in H. From [9, Theorem 4.2] , H splits over N. Let G be a complement to N in H. We will show that G is a.c. in $L(\underline{\underline{F}}_\pi \cap \underline{\underline{S}})$.

Let S be a system of finitely many equations with coefficients $\bar{g} = (g_1, \dots, g_m)$ in G that has a solution $\bar{h}$ in some $L(\underline{\underline{F}}_\pi \cap \underline{\underline{S}})$-group $G^* \geq G$. Let $\theta: H \longrightarrow G^*$ be the composition of the canonical epimorphism $\phi: H \longrightarrow G$ and $id: G \longrightarrow G^*$. Note that $N = \operatorname{Ker} \theta$. Choose an ascending chain $\{H_n\}_{n \in \mathbb{N}}$ of finite subgroups of H with

$$H_1 = \langle g_1, \dots, g_m \rangle \qquad \text{and} \qquad H = \bigcup_{n \in \mathbb{N}} H_n \ .$$

From (4.3.1), a standard embedding

$$\sigma_1 : H_1 \longrightarrow (H_1 \cap N) \text{ Wr } G^* = 1 \text{ Wr } G^*$$

with respect to some countermap θ_1^* of $\theta_1 = \theta \restriction H_1$ can be extended to an embedding

$$\sigma : H \longrightarrow W = \bigcup_{n \in \mathbb{N}} (H_n \rtimes H_n) \text{ Wr } G^* \leq (H \rtimes H) \text{ Wr } G^* \ .$$

Now, $H\sigma \leq W \in L(\underline{\underline{F}}_\pi \cap \underline{\underline{S}})$; and since $\sigma_1 = id(H_1)$, the system S has the solution $\bar{h}$ in W. But H is e.c. in $L(\underline{\underline{F}}_\pi \cap \underline{\underline{S}})$, so H contains a solution of S too. Since S consists of equations only, and since the coefficients of S lie in G, the homomorphic image G of H also has a solution for S. Thus, G is a.c. in $L(\underline{\underline{F}}_\pi \cap \underline{\underline{S}})$, as desired.

References

[1] R.J.Gregorac, On permutational products of groups, J. Austral. Math. Soc. 10 (1969), 111 - 135.

[2] P.Hall, Some constructions for locally finite groups, J. London Math. Soc. 34 (1959), 305 - 319.

[3] P.Hall, On the embedding of a group in a join of given groups, J. Austral. Math. Soc. 17 (1974), 434 - 495.

[4] K.Hickin, Universal locally finite central extensions of groups, Proc. London Math. Soc. (3) 52 (1986), 53 - 72.

[5] G.Higman, Amalgams of p-groups, J. Algebra 1 (1964), 301 - 305.

[6] J.Hirschfeld & W.H.Wheeler, Forcing, arithmetic, division rings, Springer-Verlag, Berlin 1975.

[7] O.H.Kegel & B.A.F.Wehrfritz, Locally finite groups, North Holland, Amsterdam 1973.

[8] F.Leinen, Existentially closed LX-groups, Rend. Sem. Mat. Univ. Padova 75 (1986), 191 - 226.

[9] F.Leinen, Existentially closed groups in locally finite group classes, Comm. Algebra 13 (1985), 1991 - 2024.

[10] F.Leinen, Existentially closed locally finite p-groups, J. Algebra 102 (1986), 160 - 183.

[11] F.Leinen & R.E.Phillips, Existentially closed central extensions of locally finite p-groups, Math. Proc. Camb. Phil. Soc. 100 (1986), 281 - 301.

[12] A.MacIntyre & S.Shelah, Uncountable universal locally finite groups, J. Algebra 43 (1976), 168 - 175.

[13] B.Maier, Existenziell abgeschlossene lokal endliche p-Gruppen, Arch. Math. 37 (1981), 113 - 128.

[14] B.H.Neumann, A note on algebraically closed groups, J. London Math. Soc. 27 (1952), 247 - 249.

[15] P.M.Neumann, On the structure of standard wreath products of groups, Math. Z. 84 (1964), 343 - 373.

[16] R.E.Phillips, Existentially closed locally finite central extensions; multipliers and local systems, Math. Z. 187 (1984), 383 - 392.

SOLUBLE GROUPS WITH NILPOTENT-EXTENSIBLE SUBGROUPS

John C. Lennox
Department of Pure Mathematics
University College
P.O. Box 78, Cardiff CF1 1XL
England

Suppose that $\underline{\underline{X}}$ is any class of groups. We shall say that a group G has $\underline{\underline{X}}$-extensible subgroups, and write $G \in \underline{\underline{X}}^*$, if every proper $\underline{\underline{X}}$-subgroup of G is contained in a strictly larger $\underline{\underline{X}}$-subgroup of G. Clearly $\underline{\underline{X}}$ is contained in $\underline{\underline{X}}^*$ and a natural question arises as to when equality occurs.

An easy Zorn's Lemma argument shows that the class of groups with abelian-extensible subgroups coincides with the class of all abelian groups. Indeed this argument can be applied to any class $\underline{\underline{X}}$ of groups which is closed under the operation of taking ascending unions.

Let $\underline{\underline{N}}$ be the class of all nilpotent groups. It is evident that any group with non-nilpotent hypercentre, for example a restricted direct product of nilpotent groups of class n for $n = 1,2,\ldots$, is in $\underline{\underline{N}}^*$ and so $\underline{\underline{N}}^*$ does not coincide with $\underline{\underline{N}}$. There are soluble, even metabelian groups of this type and so soluble groups with nilpotent-extensible subgroups are not necessarily nilpotent.

When we restrict our attention to finitely generated soluble groups the situation seems far less clear and so we ask

QUESTION 1. Is a finitely generated soluble group with nilpotent-extensible subgroups nilpotent?

A negative answer to this question would follow at once from a positive answer to

QUESTION 2. Does there exist a finitely generated soluble group with non-nilpotent hypercentre?

We note at this point that a nilpotent by poycyclic group with nilpotent-extensible subgroups is obviously nilpotent and it follows at once that Question 1 has an affirmative answer for finitely generated soluble linear or finitely generated soluble minimax groups since such groups are well known to be nilpotent by abelian by finite.

The property of having nilpotent-extensible subgroups is not closed under the operation of taking subgroups and this leads to

QUESTION 3. Is a finitely generated soluble group nilpotent if it and all of its finitely generated subgroups have nilpotent-extensible subgroups?

Let G be such a group and suppose that N is a nilpotent normal subgroup of G. Let $a \in N$, $g \in G$. Then $\langle a,g \rangle$ is a nilpotent by abelian $\underline{\underline{N}}^*$-group and so is nilpotent. It follows that the repeated commutator $[a,g,\dots,g]$ is eventually trivial and so a is a right Engel element of G. By a theorem of Brookes [1] this means that a belongs to the hypercentre of G. Hence N is contained in the hypercentre of G. We can now deduce

THEOREM 1. Suppose that G is a soluble group satisfying the maximal condition for normal subgroups and that G and all of its two generator subgroups are nilpotent-extensible. Then G is nilpotent.

PROOF. Let A be the last non-trivial term of the derived series of G. Then by the above argument A is contained in the hypercentre Z of G. Since G satisfies the maximal condition on normal subgroups, Z is a finite term of the upper central series of G and it readily follows that G/Z satisfies the same hypotheses as G. But G/Z has derived length less than that of G and induction completes the proof.

We turn now to consider the class $\underline{\underline{Y}}$ of finitely generated nilpotent groups. In [2] P. Hall showed that there exist finitely generated centre by metabelian groups with centre elementary abelian of countably infinite rank. If G is such a group and H is a finitely generated nilpotent subgroup of G then H cannot contain the centre of G so that there exists an element z in the centre of G and not in H. The subgroup generated by H and z is finitely generated, nilpotent and properly contains H. Thus G is in $\underline{\underline{Y}}^*$ and not in $\underline{\underline{Y}}$.

In terms of the Hall dichotomies in the class of all finitely generated soluble groups this fact leads logically to the consideration of the class of all finitely generated abelian by nilpotent groups and here we have

THEOREM 2. Suppose that G is a finitely generated abelian by nilpotent group and that G has finitely generated nilpotent-extensible subgroups. Then G is nilpotent.

The proof depends on a result on fixed points for modules over finitely generated nilpotent groups established by the author in [3].

PROOF. Suppose that G is as described in the Theorem, but G is not nilpotent. Let A be an abelian normal subgroup of G such that G/A is nilpotent and suppose x is an element of G. Then there exists by hypothesis a properly ascending sequence

$$\langle x \rangle = X_0 < X_1 < \ldots < X_n < \ldots$$

of finitely generated nilpotent subgroups X_n of G. Since G/A satisfies the maximal condition for subgroups, there exists a positive integer m such that

$$X_m A = X_{m+r} A \quad \text{for} \quad r > 1.$$

It follows that $X_m \cap A < X_{m+1} \cap A$, so that the latter subgroup, which we now denote by B, is non-trivial. Since X_{n+1} is nilpotent and B is normal there is a non-trivial element b of B in the centre of X_{n+1}. Clearly b centralizes x and so the centralizer $C_A(x)$ of x in A is non-trivial.

Regarding A as a Γ-module in the obvious way, where $\Gamma = G/A$ and G acts by conjugation we thus have that each element of Γ has a non-trivial fixed point in A. This is precisely the situation studied in [3] and it follows by the main result of that paper that some subgroup of finite index in Γ has a non-trivial fixed point in A. In particular there is a subgroup H of finite index in G which has a non-trivial centre. By coming down to a subgroup of finite index in G, if necessary, we may assume that H is characteristic in G. Hence the centre Z of H is also characteristic in G.

Using the fact that H has finite index in G it is easy to see that H is in $\underline{\underline{Y}}^*$ but not in $\underline{\underline{Y}}$ and it is then immediate that H/Z is again in $\underline{\underline{Y}}^*$ but not in $\underline{\underline{Y}}$.

We now repeat the process with H/Z in place of G and it is readily seen that this will lead to finding characteristic subgroups H_1 and Z_1 of G with Z properly contained in Z_1 such that H_1/Z_1 is in $\underline{\underline{Y}}^*$ but not in $\underline{\underline{Y}}$. Continuing in this way leads to an infinite properly ascending chain of characteristic subgroups of G which stands in contradiction to the fact that G satisfies the maximal condition for normal subgroups by [2]. This contradiction completes the proof.

REFERENCES

[1] Brookes, C. J. B.: Engel elements of soluble groups. Bull. London Math. Soc. 18 (1986), 7 - 10.

[2] Hall, P.: Finiteness conditions for soluble groups. Proc. London Math. Soc. (3) 4 (1954), 419 - 436.

[3] Lennox, J. C.: A fixed point theorem for modules over finitely generated nilpotent groups. Bull. London Math. Soc. 16 (1984), 289 - 291.

ON THE NILPOTENCE OF GROUPS WITH A CERTAIN LATTICE OF NORMAL SUBGROUPS

Patrizia Longobardi and Mercede Maj
Dipartimento di Matematica e
Applicazioni "R. Caccioppoli"
via Mezzocannone,8 - 80134 Napoli - Italy

1. Introduction

The set n(G) of the normal subgroups of a group G is a complete modular sublattice of the lattice l(G) of the subgroups of G; sometimes, if there is an isomorphism between n(G) and n(H), then one can get information about the structure of H, connected with some properties of G (see for instance [3], [4], [6], [7] and [9]).
In this direction, F. de Giovanni and S. Franciosi have generalized some previous results (see [7] and [8]), obtaining in [5] the following:

THEOREM A. *Let* G *be a nilpotent p-group of class* 2, H *be a soluble group. If* $n(G) \simeq n(H)$, *then* H *is a hypercentral p-group*.

More in general, R. Brandl has recently proved in [1] that:

THEOREM B. *Let* G *be a nilpotent p-group of class* 2, H *be a group with* $n(G) \overset{\psi}{\simeq} n(H)$.
i) *If* Z(G) *is not locally cyclic, then* H *is a hypercentral p-group*.
ii) *If* Z(G) *is locally cyclic and* N *is the only minimal normal subgroup of* G, *then* H/N^{ψ} *is a hypercentral p-group*.

In [1] R. Brandl has also asked if the hypothesis of the two above-mentioned theorems imply the nilpotence of H (of H/N^{ψ}) and even a limit to its nilpotence class.
Now we give a positive answer to this problem by proving the following:

THEOREM. *Let* G *be a nilpotent p-group of class* 2 *and* H *be a soluble group. If* $n(G) \simeq n(H)$, *then* H *is a nilpotent p-group of class* $\leqslant 3$ *and, in particular, of class* 2, *if* G' *has infinite exponent*.

There are analogous conclusions also in the hypothesis of Theorem B (see Corollaries 2.3 and 3.4).
Finally we would just like to point out that the limit to the nilpotence class of H is the best possible, as easy examples show (see [7] Bei-

spiel 2 , and Example 3.5).

Notation and terminology are the ones usually used in group theory (see for instance [10]); in particular, $Z(G)$ is the centre of the group G and $C_G(H)$ is the centralizer of H in G, where $H \leq G$.

In the sequel, we will often make use of the following straightforward Lemmas:

LEMMA 1.1. Let G be a p-group and H be a group with $n(G) \overset{\psi}{\simeq} n(H)$. If $\langle c\rangle \langle d\rangle \leq Z(G)$, $|c| \leqslant |d|$, then $C_H(\langle d\rangle^{\psi}) = C_H(\langle c\rangle^{\psi}\langle d\rangle^{\psi})$.

LEMMA 1.2. Let G be a nilpotent p-group of class 2 with $|G'| = p^k$, and H be a soluble group with $n(G) \simeq n(H)$.
If there exists $a \in Z(H)$ such that $|aH'| \geqslant p^k$, then $H/\langle a\rangle \cap H'$ is nilpotent of class $\leqslant 2$.

2. Groups of class 2 whose commutator subgroup has infinite exponent

We will prove the following

THEOREM 2.1. Let G be a nilpotent p-group of class 2 and H be a soluble group with $n(G) \overset{\psi}{\simeq} n(H)$. Assume that G' has infinite exponent. Then H is nilpotent of class 2.

First we start proving that:

2.2. Let G be a nilpotent p-group of class 2 and H be a metabelian group with $n(G) \overset{\psi}{\simeq} n(H)$. Assume that G' has infinite exponent. Then H is nilpotent of class 2.

Proof. Let K_i denote the subgroup $\Omega_i(G')$ for every $i \in \mathbb{N}$; then $G' = \bigcup_{i\in\mathbb{N}} K_i \leq Z(G)$ and $K_i \leq K_{i+1}$, for all $i \in \mathbb{N}$.
This yields $K_i^{\psi} \leq K_{i+1}^{\psi}$, $H' = (G')^{\psi}$ (see [8]) and $H' = \bigcup_{i\in\mathbb{N}} K_i^{\psi}$.
H is a hypercentral p-group (see Theorem A); thus it easily follows $K_i^{\psi} \leq Z_i(H)$ and $\exp K_i^{\psi} \leqslant p^i$, for all $i \in \mathbb{N}$.

First notice that:

(α) if $g \in C_H(y)$, where $y \in K_s^{\psi} - K_h^{\psi}$ $(s > h)$, then $g \in C_H(K_h^{\psi})$.

Let $y \in C_H(g)$. Since H' is abelian, $y \in C_H(\langle g\rangle^H)$ and hence $\langle y\rangle^H \leq$ $\leq C_H(\langle g\rangle^H)$.
Now consider $a \in K_h$, $c \in (\langle y\rangle^H)^{\psi^{-1}}$, $|c| = p^{h+1}$.
We get $\langle c\rangle \langle a\rangle = \langle c\rangle \times \langle b\rangle \leq Z(G)$, where $|b| \leqslant |a|$; by $g \in C_H(\langle c\rangle^{\psi})$ and by Lemma 1.1 , we obtain $g \in C_H(\langle a\rangle^{\psi})$.

The required conclusion follows from $K_h = \bigvee_{a \in K_h} \langle a \rangle$.

Next we claim that:

(β) <u>if</u> $H' \leq S \leq H$ <u>and</u> S/H' <u>is finite</u>, <u>then</u> $S^{\psi^{-1}} = G'C$, <u>where</u> $C \triangleleft G$, $G' \cap C \leq K_n$ <u>for a suitable</u> $n \in \mathbb{N}$, <u>and</u> S <u>is nilpotent</u>.

By hypothesis, S is normal in H and $S^{\psi^{-1}}/G'$ is finite, whence $S^{\psi^{-1}} = G'X$ where X is finite. Suppose, for instance, $|X| = p^n$, and write C for X^G. Since $G' \leq Z(G)$, it easily follows that $\exp C \leqslant p^n$, so that $G' \cap C \leq K_n$. Then $S^{\psi^{-1}} = G'C$ implies $S = H'C^{\psi}$, where $C^{\psi} \cap H' \leq K_n^{\psi} \leq Z_n(H)$. Finally S is nilpotent, since $S/C^{\psi} \cap H'$ is abelian.

Now, by contradiction, let $H' \not\leq Z(H)$. Then there exist $h \in \mathbb{N}$ and $y \in H$ such that $y \notin C_H(K_h^{\psi})$. By (β) $H'\langle y \rangle$ is nilpotent and by (α) $Z(H'\langle y \rangle) \cap H' \leq K_h^{\psi}$, so that $Z(H'\langle y \rangle)$ and hence $H' \leq H'\langle y \rangle$ has finite exponent.

Put $\exp H' = p^k$.

If a is an element of G such that aG' has order p^{k+1}, then $G'\langle a \rangle = G'C$, where $C \triangleleft G$, $G' \cap C \leq K_{k+1}$ (see (β)). This yields $\frac{G'C}{G' \cap C} = \frac{G'}{G' \cap C} \times \frac{C}{G' \cap C}$, where $\frac{C}{G' \cap C} \leq Z(\frac{G}{G' \cap C})$, $\frac{C}{G' \cap C}$ cyclic of order p^{k+1} and $\exp \frac{G'}{G' \cap C}$ infinite.

Write now $\Omega_{k+1}(\frac{G'}{G' \cap C}) = \frac{T}{G' \cap C}$; Lemma 1.1 shows that

$$C_{H/H' \cap C^{\psi}}\left(\frac{C^{\psi}}{(G' \cap C)^{\psi}}\right) \leq C_{H/H' \cap C^{\psi}}\left(\frac{T}{(G' \cap C)^{\psi}}\right).$$

We also have $\frac{H'}{(G' \cap C)^{\psi}} \cap \frac{C^{\psi}}{(G' \cap C)^{\psi}} = \frac{(G')^{\psi} \cap C^{\psi}}{(G' \cap C)^{\psi}} = 1$, so that $\frac{C^{\psi}}{(G' \cap C)^{\psi}} \leq Z(\frac{H}{(G' \cap C)^{\psi}})$ and hence $\frac{T^{\psi}}{(G' \cap C)^{\psi}} \leq Z(\frac{H}{(G' \cap C)^{\psi}})$. Thus ψ induces an isomorphism between the lattices $\mathfrak{l}(\frac{T}{G' \cap C})$ and $\mathfrak{l}(\frac{T^{\psi}}{(G' \cap C)^{\psi}})$ and we can easily see that $\frac{T}{G' \cap C} \simeq \frac{T^{\psi}}{(G' \cap C)^{\psi}}$. Since $\exp \frac{T}{G' \cap C} = p^{k+1}$, it follows $\exp \frac{T^{\psi}}{(G' \cap C)^{\psi}} = p^{k+1}$, contradicting $\frac{T^{\psi}}{(G' \cap C)^{\psi}} \leq \frac{H'}{(G' \cap C)^{\psi}}$ and $\exp H' = p^k$. $\triangle$

We are now able to prove the above-mentioned Theorem 2.1.

<u>Proof of</u> THEOREM 2.1. By Theorem A, H is a hypercentral p-group.

As in 2.2, let $K_i = \Omega_i(G')$ for all $i \in \mathbb{N}$ and write $K_i = D_i \times (\times_{j \in J_i} C_{ij})$,

where $D_i \leq K_{i-1}$, $C_{ij} = <c_{ij}>$, $|C_{ij}| = p^i$, for all $i \in \mathbb{N}$.

Arguing as in 2.2, we get $H' = (G')^{\Psi} = \bigcup_{i \in \mathbb{N}} K_i^{\Psi}$, $K_i^{\Psi} \leq Z_i(H)$ and exp $K_i^{\Psi} \leqslant$ $\leqslant p^i$, for every $i \in \mathbb{N}$.

First suppose $|J_i| > 1$, for all $i \in \mathbb{N}$; if $j, k \in J_i$, $j \neq k$, then $C_{ij}^{\Psi} \leq$ $\leq C_H(C_{ik}^{\Psi})$ and $C_H(K_i^{\Psi}) = C_H(C_{ik}^{\Psi})$ (see Lemma 1.1). Thus each C_{ij}^{Ψ} is abelian.

Analogously, from $D^{\Psi} \leq C_H(C_{ik}^{\Psi}) = C_H(K_i^{\Psi})$ it follows D^{Ψ} abelian; so K_i^{Ψ} is abelian, for all $i \in \mathbb{N}$. Since $H' = \bigcup_{i \in \mathbb{N}} K_i^{\Psi}$, where $K_i^{\Psi} \leq K_{i+1}^{\Psi}$, we obtain $H'' = 1$.

Otherwise, suppose that there exists $i \in \mathbb{N}$ such that $|J_i| = 1$.

We have $|K_i/K_{i+1}| = p$, so that $|\Omega_1((G')^{p^{i-1}})| = p$ and $(G')^{p^{i-1}} \simeq Z(p^{\infty})$.

It follows $G' = (G')^{p^{i-1}} \times T$, with $T^{p^{i-1}} = 1$, which yields $H' =$ $= ((G')^{p^{i-1}})^{\Psi} \times T^{\Psi}$, where $T^{\Psi} \leq K_{i-1}^{\Psi}$ and $C^{\Psi^{-1}}$ finite for every $C \lhd H$, $C <$ $< ((G')^{p^{i-1}})^{\Psi}$.

Denoting $(((G')^{p^{i-1}})^{\Psi})'$ by D, we obtain $D \lhd H$ and $H'' = D \times (T^{\Psi})'$; so $(H'')^{\Psi^{-1}} = L \times ((T^{\Psi})')^{\Psi^{-1}}$ where L is finite; this implies $(H'')^{\Psi^{-1}} \leq K_n$ for a suitable $n \in \mathbb{N}$ and $H'' \leq K_n^{\Psi} \leq Z_n(H)$. By 2.2, H/H'' is nilpotent of class $\leqslant 2$ and hence H is nilpotent with $[H', H] \leq H''$. But $H'' \leq [H', H, H]$, which yields $[H', H] = [H', H, H]$ and H nilpotent of class 2. $\triangle$

In a very similar way we can get the following

COROLLARY 2.3. Let G be a nilpotent p-group of class 2 and H be a group with $n(G) \overset{\Psi}{\simeq} n(H)$. Assume that G' has infinite exponent. Then: (i) if Z(G) is not locally cyclic, then H is nilpotent of class 2; (ii) if Z(G) is locally cyclic and if N denotes the unique minimal normal subgroup of G, then H/N^{Ψ} is a nilpotent p-group of class 2.

Proof. (i) By Theorem B, H is a hypercentral p-group; then $A' < A$ for every $A \leq H$. Arguing as in the previous proof, we can get that H is nilpotent of class 2.

(ii) It easily follows from Theorem B, Theorem 2.1 and (i). $\triangle$

3. Nilpotent groups of class 2, whose commutator subgroup has finite

exponent

We will get the following

THEOREM 3.1. *Let* G *be a nilpotent p-group of class* 2 *and* H *be a soluble group with* $n(G)\overset{\psi}{\simeq}n(H)$. *Assume that* G' *has finite exponent. Then* H *is a nilpotent p-group of class* $\leqslant 3$.

To prove this we will need two auxiliary results:

3.2. *Let* G *be a nilpotent p-group of class* 2 *and* H *be a hypercentral p-group with* $n(G)\overset{\psi}{\simeq}n(H)$. *Assume that* G' *has finite exponent* p^t. *Then* H *is a nilpotent p-group of class* $\leqslant t+1$.

Proof. Denoting, for every $i\in\mathbb{N}$, $\Omega_i(G')$ by K_i, we get $K_i\leq K_{i+1}$ for all $i\in\mathbb{N}$, and $G'=K_t$.
From H locally nilpotent, it easily follows that $K_i^{\psi}\leq Z_i(H)$ for all $i\in\mathbb{N}$; then $H'\leq Z_t(H)$, since $H'=(G')^{\psi}\leq K_t^{\psi}$ (see [8]). Therefore H is nilpotent of class $\leqslant t+1$. △

3.3. *Let* G *be a nilpotent p-group of class* 2 *with* G' *cyclic of order* $\geqslant p^3$. *If* H *is a soluble group with* $n(G)\overset{\psi}{\simeq}n(H)$, *then* H' *is cyclic*.

Proof. By Theorem A and 3.2, H is a nilpotent p-group.
Put $|G'|=p^k$, it can easily be proved that $|H'|=|(G')^{\psi}|=p^k$.

Since $|G'|\geqslant p^3$, there exist in $\frac{G}{G'}\simeq\frac{H}{H'}$ two independent elements of order p^k.

Assume, by contradiction, H' not cyclic.
$H'/\Phi(H')$ is an elementary abelian p-group of order $>p$.

Let $G'=\langle c\rangle$. Then $\Phi(H')^{\psi^{-1}}\leq\langle c^{p^2}\rangle$ and $H'/\langle c^{p^3}\rangle^{\psi}$ is not cyclic.

We can assume $|G'|=|H'|=p^3$.
Write $\bar{G}=\dfrac{G}{\langle c^{p^2}\rangle}$, $\bar{H}=\dfrac{H}{\langle c^{p^2}\rangle^{\psi}}$.
Since $n(\bar{G})\simeq n(\bar{H})$ and $\bar{H}'$ is an elementary abelian p-group of order p^2, it follows that $\bar{H}$ is of class >2; then $\bar{H}$ is of class 3 by 3.2.
Moreover $|\bar{H}:C_{\bar{H}}(\bar{H}')|=p$, so that there exists $\overline{xH'}\in C_{\bar{H}}(\bar{H}')$ with $|\overline{xH'}|=p^3$. This yields $\bar{x}^p\in Z(\bar{H})$, $x^{p^2}\in Z(H)$ and by Lemma 1.2, $\langle\bar{x}^p\rangle\cap\bar{H}'\neq 1$.
Then we have $\dfrac{(\langle x^p\rangle\cap H')\langle c^{p^2}\rangle^{\psi}}{\langle c^{p^2}\rangle^{\psi}}\neq 1$ and $\langle x^p\rangle\cap H'=\langle x^{p^2}\rangle\cap H'\lhd H$;

so $\langle c^{p^2}\rangle^{\psi}<\langle x^{p^2}\rangle\cap H'$, since $n(G')$ is a chain.

Therefore H' is abelian and n(G') chain implies $\Omega_1(H') \leq \langle x^{p^2} \rangle \cap H'$. Consequently H' is cyclic, a contradiction. △

Now we are able to prove Theorem 3.1.

Proof of THEOREM 3.1. G' has finite exponent, so we can assume G' cyclic. If $|G'| = p^2$, then H is of class $\leqslant 3$ by 3.2.
Suppose now $p^k = |G'| \geqslant p^3$. By 3.3 H' is cyclic; write $H' = \langle a \rangle$.
Obviously $|H'| = |G'| = p^k \geqslant p^3$.

First suppose $p > 2$ or $p = 2$ and $[H', H] \leq \langle a^4 \rangle$, and put $\bar{H} = \frac{H}{Z(H)}$ and $\bar{G} = \frac{G}{(Z(H))^{\psi^{-1}}}$.

Arguing as in [2], Lemma 3, we obtain $\bar{H} = \langle x, y \rangle M$, with $M \leq Z(\bar{H})$, $y^x = y^{1+p^s}$, $\exp M \leqslant p^s$.

This implies $\bar{H}' = \langle y^{p^s} \rangle$, $\frac{\bar{H}}{\bar{H}'} = \langle \overline{xH'} \rangle \frac{N}{\bar{H}'}$, where $\exp \frac{N}{\bar{H}'} \leqslant p^s$.

If $\bar{G}$ is locally cyclic, then $G' \leq Z(H)^{\psi^{-1}}$ yields $H' = (G')^{\psi} \leq Z(H)$ and H is of class 2, as required.

If $\bar{G}$ is not locally cyclic, we obtain (see [8]) $\frac{\bar{G}}{\bar{G}'} \simeq \frac{\bar{H}}{\bar{H}'}$, and therefore

$\frac{\bar{G}}{\bar{G}'} \simeq \langle x\bar{H}' \rangle \frac{N}{\bar{H}'}$, where $\exp \frac{N}{\bar{H}'} \leqslant p^s$.
From $\bar{G}' \leq Z(\bar{G}')$ it easily follows $|\bar{G}'| \leqslant p^s$, and so $|\bar{H}'| \leqslant p^s$, which implies $\bar{H}' \leq Z(\bar{H})$ and H of class $\leqslant 3$.

Finally, let $p = 2$ and $[H', H] = \langle a^2 \rangle$ and assume, by contradiction, $|a| \geqslant 8$. Put $\bar{H} = \frac{H}{\langle a^8 \rangle}$ and $\bar{G} = \frac{G}{(\langle a^8 \rangle)^{\psi^{-1}}}$, there exists $x \in H$ such that $|xH'| \geqslant 8$ and $\bar{a}^x = \bar{a}^{-1+2k}$ for a suitable odd k.

Then $\langle \bar{a} \rangle \cap \langle \bar{x} \rangle \leq \langle \bar{a}^4 \rangle$ and $\langle \bar{x}^2, \bar{a}^4 \rangle \lhd \bar{H}$ and hence $\bar{H}/\langle \bar{a}^4 \rangle$ is of class $\leqslant 2$ (see Lemma 1.2), a contradiction. △

COROLLARY 3.4. Let G be a nilpotent p-group of class 2 and H be a group with $n(G) \overset{\psi}{\simeq} n(H)$. Assume G' with finite exponent. Then:
(i) if Z(G) is not locally cyclic, then H is nilpotent of class $\leqslant 3$;
(ii) if Z(G) is locally cyclic, then H/N^{ψ} is nilpotent of class $\leqslant 3$, where N is the unique minimal normal subgroup of G.

Proof. (i) By Theorem B, H is a hypercentral p-group; whence H is nilpotent by 3.2 and the result follows from Theorem 3.1.

(ii) As before. $\triangle$

Finally, we would like to remark that there exist groups G and H satisfying the hypothesis of Corollary 3.4 (i), where H is nilpotent of class 3, as shown by the following

EXAMPLE 3.5. Let $G = ((\langle a\rangle \times \langle c\rangle) \rtimes \langle b\rangle) \times \langle d\rangle$, where $a^{p^2} = b^{p^2} = c^{p^2} = d^p = 1$, $[a, b] = c$, $[b, c] = 1$, and let $H = ((\langle x\rangle \times \langle z\rangle) \rtimes \langle y\rangle) \times \langle t\rangle$, where $x^{p^2} = y^{p^2} = z^{p^2} = t^p = 1$, $[x, y] = z$, $[z, y] = z^p$, p odd prime.

Then $n(G) \simeq n(H)$, G is of class 2 and H of class 3.

Proof. We have $Z(G) \geq A = \langle c^p\rangle \times \langle d\rangle$, $Z(H) = \langle z^p\rangle \times \langle t\rangle$, $G' = \langle c\rangle$, $H' = \langle z\rangle$, $A \simeq Z(H)$ and $\frac{G}{\langle c^p\rangle} \simeq \frac{H}{\langle z^p\rangle}$.

Now let $f: A \to Z(H)$ be the isomorphism induced by $f(c^p) = z^p$, $f(d) = t$, and let φ be an isomorphism between $n(\frac{G}{\langle c^p\rangle})$ and $n(\frac{H}{\langle z^p\rangle})$ such that $(\frac{A}{\langle c^p\rangle})^{\varphi} = \frac{Z(H)}{\langle z^p\rangle}$.

Finally, define $\psi : n(G) \to n(H)$ by:

$$\frac{N^{\psi}}{\langle z^p\rangle} = (\frac{N}{\langle c^p\rangle})^{\varphi}, \text{ if } N \in n(G) \text{ and } N \geq \langle c^p\rangle,$$

$$N^{\psi} = f(N), \text{ if } N \in n(G) \text{ and } N \not\geq \langle c^p\rangle.$$

It can be easily shown that ψ is an isomorphism; obviously G is of class 2 and H of class 3, as required. $\triangle$

References

[1] R. Brandl - On groups with certain lattices of normal subgroups, Arch. Math., to appear.

[2] Y. Cheng - On finite p-groups with cyclic commutator subgroup, Arch. Math., 39 (1982), 295-298.

[3] M. Curzio - Una caratterizzazione reticolare dei gruppi abeliani, Rend. Mat., 24 (1965), 1-10.

[4] A. Franchetta - F. Tuccillo - Su una classe di gruppi ipercentrali, Atti Accad. Naz. Lincei, 59 (1975), 232-237.

[5] F. de Giovanni - S. Franciosi - Alcuni epimorfismi tra reticoli di sottogruppi normali, Ist. Lombardo (Rend. Sc.), A 116 (1982),

45 - 53.

[6] F. de Giovanni - S. Franciosi - Isomorfismi tra reticoli di sottogruppi normali di gruppi nilpotenti senza torsione, Ann. Univ. Ferrara Sc. Mat., 91 (1985), 91-98.

[7] H. Heineken - Über die Charakterisierung von Gruppen durch gewisse Untergruppenverbände, J. Reine Angew. Math., 220 (1965), 30-36.

[8] P. Longobardi - M. Maj - Su di un teorema di Heineken, Riv. Mat. Univ. Parma, 4 (1976), 315-320.

[9] P. Longobardi - M. Maj - Su alcuni gruppi con il reticolo dei sottogruppi normali isomorfo al reticolo dei sottogruppi normali di un prodotto libero, Rend. Mat., 3 (1983), 725-734.

[10] D. J. S. Robinson - A Course in the Theory of Groups, Springer-Verlag, Berlin 1980.

[11] M. Suzuki - Structure of a group and the structure of its lattice of subgroups, Springer-Verlag, Berlin 1965.

TORSION-FREE NILPOTENT GROUPS WITH BOUNDED RANKS OF THE ABELIAN SUBGROUPS

Walter Möhres
Universität Würzburg

If G is a group and $1=G_0\leq\ldots\leq G_n=G$ is a subnormal series with each factor a torsion group or torsion-free abelian of rank 1, let H(G) be the number of nontrivial torsion-free factors. H(G) is of course independent of the choice of the series. We shall say $H(G)=\infty$ if G has no such series. As in the polycyclic case we call H(G) the Hirsch-length of G.

Moreover for any group G let a(G) be the maximal Hirsch-length of the abelian subgroups and n(G) that of the normal abelian subgroups. By T we will denote the class of torsion-free nilpotent groups G with $a(G)<\infty$.

Theorems 1 and 2 will show how the nilpotency class c(G) and the Hirsch-length H(G) of a group $G\in T$ are bounded in terms of a(G) or n(G). Theorem 3 tells us that a(G) and n(G) are equal for groups G of low class. At the end the lower and upper central series of groups $G\in T$ with $a(G)\leq 4$ will be investigated.

THEOREM 1: For every $n\in\mathbb{N}$ we have

$$\max\{c(G)\mid G\in T,\ a(G)=n\} = \max\{c(G)\mid G\in T,\ n(G)=n\} = \begin{cases} 2n-1 & \text{if } n\neq 2 \\ 2 & \text{if } n=2 \end{cases}$$

To prove this Theorem we need the following

LEMMA: If $G\in T$, $G\neq 1$, then $c(G)\leq 2n(G)-1$.

PROOF: Let $n=n(G)$, $c=c(G)$. There is a positive integer m such that $c+2\geq 2m\geq c+1$. If $K_i(G)$ is the i-th term of the lower central series of G, we have $K_m(G)'\leq K_{2m}(G)\leq K_{c+1}(G)=1$. Thus $K_m(G)$ is abelian and $H(K_m(G))\leq n$. Furthermore for every $s\in\mathbb{N}$ there is an epimorphism $G/G'\otimes K_s(G)/K_{s+1}(G)\rightarrow K_{s+1}(G)/K_{s+2}(G)$, so $K_s(G)/K_{s+1}(G)$ cannot be a torsion group for $s\leq c$. Hence we have $H(K_m(G))\geq c+1-m\geq c/2$ and $c\leq 2n$.

Assume $c=2n$. Then $K_{n+1}(G)$ is abelian and $n\geq H(K_{n+1}(G))\geq c+1-(n+1)=n$. Let I be the isolator of $K_{n+1}(G)$ in G. Then I is a normal abelian subgroup of G and it has to be maximal with this property, because $H(I)=n$ and G/I is torsion-free. Thus $I=C(I)$, the centralizer of I in G. But $C(I)=C(K_{n+1}(G))$, hence

$I=C(K_{n+1}(G))$. Because of $[K_n(G), K_{n+1}(G)] \leq K_{2n+1}(G) = 1$ we have $K_n(G) \leq C(K_{n+1}(G))=I$ and $K_n(G)/K_{n+1}(G)$ is a torsion group. This is a contradiction.

The Lemma provides half of the proof of Theorem 1. The case $n \leq 2$ is trivial. So it remains to construct for every $n \geq 3$ a group $G_n \in T$ with $a(G_n)=n$ and $c(G_n)=2n-1$ (see [1], p.14).

THEOREM 2: If $G \in T$, then $H(G) \leq n(G)(n(G)+1)/2 \leq a(G)(a(G)+1)/2$.

PROOF: Let A be a maximal normal abelian subgroup of G and let $m=H(A)$. Then $C(A)=A$ and thus G/A can be embedded in $GL(m,\mathbb{Q})$. Moreover, because G is nilpotent, G/A can be embedded in $U(m,\mathbb{Q})$, the group of unitriangular matrices of dimension m over $\mathbb{Q}$. Now $H(U(m,\mathbb{Q}))=m(m-1)/2$ and hence $H(G)=H(A)+H(G/A)\leq m+m(m-1)/2=m(m+1)/2$. The statement follows because $m \leq n(G)$.

THEOREM 3: If $G \in T$ with $c(G) \leq 4$, then $a(G)=n(G)$.

Here we will only give the proof for $c(G)\leq 3$. The case $c(G)=4$ may be handled by similar methods (see [1], p.19).

PROOF for $c(G)\leq 3$: Let G be a counterexample with $H(G)$ minimal. Let U be an abelian subgroup of G with $H(U)=a(G)$ and containing the centre $Z(G)$ of G. There exists a subgroup N of G containing U with $H(N)=H(G)-1$. By minimality of $H(G)$ there is a normal abelian subgroup V of N with $H(V)=H(U)=a(G)$ and we may choose V such that it contains $Z(G)$. Let $a \in G\setminus I(N)$ and $W=\{v \in V \mid [v,a] \in V\}$. Then W is a subgroup because $K_3(G)$ is contained in V. Let $X=\langle W,[V,a]\rangle$.

By $Z_i(G)$ we will denote the terms of the upper central series of G. If $x \in N$, then $[W,x]\leq [V,x]\leq V \cap Z_2(G)\leq W \leq X$ and $[V,a,x]\leq Z(G)\leq W \leq X$. Moreover $[X,a]\leq W \leq X$. Hence $\langle a,N\rangle \leq N(X)$, where $N(X)$ is the normalizer of X in G, and $G=I(\langle a,N\rangle)\leq I(N(X))\leq N(I(X))$, so $I(X)$ is normal in G.

Let $v \in V$, $w \in W$. Then $[w,a] \in V$ and $[w,a,v]=1$. Because $c(G)\leq 3$ we have by the Hall-Witt-identity $1=[w,a,v][a,v,w][v,w,a]=[a,v,w]=[v,a,w]^{-1}$. Hence X is abelian.

$Z(G)$ is contained in W and so we have a homomorphism $f:V \to X/W$ defined by $f(v)=[v,a]W$ for $v \in V$. If K is the kernel of f, then $K \leq W$. Thus $a(G)=H(V)=H(V/K)+H(K)\leq H(X/W)+H(W)=H(X)$.

Now $I(X)$ is a normal abelian subgroup of G with $H(I(X))=H(X)\geq a(G)$, a contradiction.

For every group $G \in T$ with $c=c(G)$ we have the following vectors of positive integers:

$$\gamma(G)= (H(K_c(G)/K_{c+1}(G)),\dots,H(K_1(G)/K_2(G)))$$
$$\zeta(G)= (H(Z_1(G)/Z_0(G)),\dots,H(Z_c(G)/Z_{c-1}(G)))$$

I have calculated the possible values of $(\gamma(G), \zeta(G))$ for groups $G \in T$, if $a(G) \leq 4$. The following table will give a survey:

a(G)	1	2	3	4
number of possibilities for $(\gamma(G), \zeta(G))$	1	2	8	53
maximal H(G)	1	3	6	9

The proof of that and an example of every possible type will be found in [1].

References

[1] W.Möhres, Torsionsfreie nilpotente Gruppen mit beschränktem Rang der abelschen Untergruppen, Diplomarbeit, Würzburg 1986

On permutation properties for semigroups

Giuseppe Pirillo

Summary. We present a semigroup which does not have the permutation property P *defined by Restivo and Reutenauer in* [3], *but satisfies a weaker permutation property* P^*.

In [3] Restivo and Reutenauer introduced the following definition.

Definition 1. Let S be a semigroup and let n be an integer $\geqslant 2$. We say that S has the property P_n iff, for every n-tuple of elements

$$x_1, x_2, \ldots, x_n$$

in S, there is some permutation φ of $\{1, 2, \ldots, n\}$, $\varphi \neq id$, which satisfies

$$x_1 \cdot x_2 \cdot \ldots \cdot x_n = x_{\varphi(1)} \cdot x_{\varphi(2)} \cdot \ldots \cdot x_{\varphi(n)}.$$

We say that S has the property P iff, for some integer $n \geqslant 2$, the semigroup S has the property P_n.

The relevance of these notions in group theory is shown in [1] and [2].

Definition 2. Let S be a semigroup and let n be an integer $\geqslant 2$. We say that S has the property P_n^* iff, for every n-tuple of elements

$$x_1, x_2, \ldots, x_n$$

in S, there are two permutations φ and ψ of $\{1, 2, \ldots, n\}$, $\varphi \neq \psi$, such that

$$x_{\varphi(1)} \cdot x_{\varphi(2)} \cdot \ldots \cdot x_{\varphi(n)} = x_{\psi(1)} \cdot x_{\psi(2)} \cdot \ldots \cdot x_{\psi(n)}.$$

We say that S has the property P^* iff, for some integer $n \geqslant 2$, the semigroup S has the property P_n^*.

We show that there exists a semigroup with the property P^* and without the property P. More precisely the following proposition holds.

Proposition 1. There exists a semigroup which has the property P_3^* and which, for each $n \geqslant 2$, does not have the property P_n.

Proof. Let $\mathbb{P}$ be the set of all the positive integers, which we consider as an alphabet. Let $\leqslant$ be the usual order relation in $\mathbb{P}$. Let $\mathbb{P}^+$ be the free semigroup on $\mathbb{P}$. If $w \in \mathbb{P}^+$, we denote by $|w|$ the length of w and we denote by $w(i)$ the i-th letter of w.

Let the subset I of $\mathbb{P}^+$ be defined as follows

$$I = \{ w \in \mathbb{P}^+ / \exists\, i,j \in \mathbb{P} \text{ such that } 1 \leqslant i < j \leqslant |w| \text{ and } w(i) > w(j) \}.$$

The subset I of $\mathbb{P}^+$ is an ideal of $\mathbb{P}^+$ and the Rees quotient

$$S = \mathbb{P}^+ / I$$

has the required properties.

Indeed, to show that S has the property P_3^*, it is enough to use the definition of I. To show that, for each $n \geqslant 2$, S does not have the property P_n, it is sufficient to note that

$$1.2.\dots.n \notin I$$

and, for every $\varphi \neq id$,

$$\varphi(1).\varphi(2).\dots.\varphi(n) \in I.$$

REFERENCES

1. M. CURZIO, P. LONGOBARDI and M. MAJ, Su un problema combinatorio in teoria dei gruppi, Atti Acc. Lincei Rend. fis. VIII, 74, 136-142 (1983).
2. M. CURZIO, P. LONGOBARDI, M. MAJ and D.J.S. ROBINSON, A permutational property of groups, Arch. Math., 44, 385-389 (1985).
3. A. RESTIVO and C. REUTENAUER, On Burnside problem for semigroups, J. Algebra, 89, 102-104 (1984).

GIUSEPPE PIRILLO

I.A.G.A.-I.A.M.I.-C.N.R.

Viale Morgagni 67/A

FIRENZE (ITALIA)

Vanishing Theorems for Cohomology of Locally Nilpotent Groups

Derek J.S. Robinson
Department of Mathematics,
University of Illinois in Urbana-Champaign,
Urbana, Il. 61801, USA

1. Background

We begin with a well-known result dating back to the 1950's , due to Gaschütz [6] and Schenkman [21].

Theorem

Let E be a finite group with largest nilpotent quotient E/A. If A is abelian, then E splits over A and all complements of A in E are conjugate.

To understand the homological significance of the theorem it is necessary to recall the group theoretic interpretation of (co)-homology in low dimensions.

Let G be a group and A a $\mathbb{Z}G$-module.

(i) $H^0(G,A) = A^G$, the set of G-fixed points in A.

(ii) $H_0(G,A) = A_G = A/[A,G]$ where $[A,G]$ is the subgroup generated by all $a(g-1)$, $a \in A$, $g \in G$.

(iii) $H^1(G,A)$ corresponds bijectively to the set of conjugacy classes of complements of A in the semidirect product $G \ltimes A$.

(iv) $H^2(G,A)$ corresponds bijectively to the set of equivalence classes of extensions of A by G (inducing the given module structure in A).

The Gaschütz-Schenkman theorem may therefore be reformulated as

Theorem.

Let G be a finite nilpotent group and A a finite $\mathbb{Z}G$-module. If $H_0(G,A) = 0$, then $H^1(G,A) = 0 = H^2(G,A)$.

For finite groups very wide generalizations of this are now known which apply to saturated formations of finite groups (see [14] and [1]). Here we are concerned with generalizations in a different direction, that of infinite groups.

The next impetus to the subject came from topology. In 1975

Brown and Dror [3] and Dwyer [5] proved the following result, which has applications to the theory of prenilpotent spaces.

Theorem

Let G be a finitely generated nilpotent group and A a finitely generated (and hence noetherian) $\mathbb{Z}G$-module. If $H_0(G,A) = 0$, then $H_n(G,A) = 0$ for all $n \geq 0$.

These results suggest the existence of vanishing theorems for (co)homology of nilpotent groups:given that (co)homology is zero in low dimensions (0 or perhaps 1), must it be zero in all dimensions?

Simple examples show that some type of finiteness condition must be imposed on the module. In practice this usually takes the form of a chain condition or finiteness of rank.

The following result appeared in 1976; it generalized most of the previously published work (see [3], [4], [5], [8], [20]).

Theorem ([17])

Let G be a nilpotent group, R a ring with identity and A an RG-module. If either

(i) A is noetherian and $H_0(G,A) = 0$

or

(ii) A is artinian and $H^0(G,A) = 0$,

then $H_n(G,A) = 0 = H^n(G,A)$ for all $n \geq 0$.

As an example of an application to group theory we cite: let E be a group with the minimal condition for normal subgroups and let A be an abelian normal subgroup such that E/A is nilpotent. If $A \cap Z(G) = 1$, then E splits over A and all complements of A are conjugate. In addition several vanishing theorems for $\mathbb{Z}G$-modules with finite $\mathbb{Z}$-rank were given in [17]. Some of these have proved quite useful in infinite soluble group theory.

2. Problems

After this brief review of the theory as it existed in 1976, we shall mention some problems left open at the time, as well as others suggested by more recent work.

(i) It was left open in [17] whether the theory could be extended to locally nilpotent groups. There was already some evidence that this might be possible in the form of splitting and conjugacy theorems due to Hartley and Tomkinson [11]. Further results of this kind have since then been found by Hartley [10] and Zaicev [22], [23].

The main obstacle to such an extension is the well-known failure of the cohomology functor to commute with direct limits. (There is, of course, no such trouble with homology). There are techniques available to overcome this difficulty and some of these will be described below. The end result is that it is possible to construct a satisfactory theory for locally nilpotent groups, containing all the known splitting and conjugacy theorems.

(ii) All the finite rank theorems in [17] were proved for $\mathbb{Z}G$-modules with finite $\mathbb{Z}$-rank. It is fairly evident that the proofs carry through if $\mathbb{Z}$ is replaced by any principal ideal domain (a case which has already been found useful in group theory). There remains the problem of allowing more general types of ring R.

Here there are some natural limitations since one must have a sensible definition of rank. It would seem reasonable to ask that R be at least a noetherian domain. In practice it has only proved possible to carry the theory through in its entirety for Dedekind domains.

(iii) Among the new splitting theorems there is a result of Hartley [10] which is notable in that it allows modules with infinitely many primary components. In such a situation the difficulties presented by cohomology-versus-direct limits arise again.

It turns out that Hartley's theorem is also a special case of a vanishing theorem for locally nilpotent torsion groups. Here, however, the situation is more complex since the cardinal of the group plays a role.

In the remainder of the article we shall describe recent progress in solving the above problems, and indicate some of the techniques involved. Full details will appear in a forthcoming work [19].

3. Methods

We shall review some of the techniques that can be used to treat cohomology of groups with local systems of subgroups. The first two are elementary.

Proposition 1

Let G be a group which is the union of a complete well-ordered ascending chain of subgroups $\{G_\alpha | \alpha < \beta\}$. Let A be a $\mathbb{Z}G$-module and suppose that there is a positive integer n with the property

$H^{n-1}(G_\alpha, A) = 0 = H^n(G_\alpha, A)$ for all $\alpha < \beta$. Then $H^n(G,A) = 0$.

This is easily proved by a direct argument with cocycles. Proposition 1 is often sufficient to deal with countable groups; however, in more general situations the following result can frequently be used with effect.

Proposition 2

Let G be a group and let $\underline{S}$ be a set of subgroups of G closed under finite joins such that G is generated by the members of $\underline{S}$. Let A be a $\mathbb{Z}G$-module and suppose that there is a positive integer n such that $H^i(G,A) = 0$ for $i = 0, 1, \ldots, n$ and all S in $\underline{S}$. Then $H^i(G,A) = 0$ for $i = 0, 1, \ldots, n$.

Proof

Well-order the set $\underline{S}$ as $\{S_\alpha \mid 0 < \alpha < \gamma\}$, with γ an ordinal. Define $G_0 = 1$ and $G_\beta = \langle S_\alpha \mid \alpha < \beta \rangle$ if $0 < \beta \leq \gamma$. Then $\{G_\beta \mid 0 \leq \beta \leq \gamma\}$ is a complete well-ordered ascending chain with $G_\gamma = G$. Assuming the result to be false one can find a least ordinal α such that $H^i(\langle S, G_\alpha \rangle, A) \neq 0$ for some $i \leq n$, and $S \in \underline{S}$. Moreover $\alpha > 0$ and $i > 0$. If α is a limit ordinal, $\langle S, G_\alpha \rangle$ is the union of the chain $\langle S, G_\beta \rangle$, $\beta < \alpha$, and $H^j(\langle S, G_\beta \rangle, A) = 0$ for all $j \leq n$. Proposition 1 gives a contradiction.

Consequently α is not a limit ordinal and $G_\alpha = \langle S_{\alpha-1}, G_{\alpha-1} \rangle$. But then $T = \langle S, S_{\alpha-1} \rangle \in \underline{S}$ and

$$H^i(\langle S, G_\alpha \rangle, A) = H^i(\langle T, G_{\alpha-1} \rangle, A) = 0$$

for $i = 0, 1, \ldots, n$, a contradiction.

A spectral sequence for direct limits of groups

Let (S_i) be a direct system of groups and homomorphisms and let $G = \varinjlim(S_i)$. If A is an RG-module, R being any ring with identity, A becomes an RS_i-module via the obvious mapping $S_i \to G$. Then $(H^q(S_i, A))$ is an inverse system of R-modules for each $q \geq 0$; here the homomorphisms are the induced maps $H^q(S_j, A) \to H^q(S_i, A)$, $j \geq i$.

It is a consequence of the Grothendieck spectral sequence (see [9] or [12], p.297) that there is a cohomology spectral sequence $\{E^{pq}\}$ converging to $H^n(G,A)$; indeed

$$E_2^{pq} = \varprojlim{}^{(p)} (H^q(S_i, A)) \Rightarrow H^n(G,A)$$

where $p + q = n$. Here $\varprojlim^{(p)}$ is the pth derived functor of $\varprojlim$ (for an account of these derived functors see [15]).

This spectral sequence is a powerful tool in studying the cohomology of groups with a local system. Notice that Proposition 2 is an immediate consequence.

4. Results

We begin with a theorem on modules with chain conditions which generalizes [17], Theorems A and B, [22], Theorem 4 and [23], Theorem 2.

Let G be a group, R a ring with identity, N a normal subgroup and A an RG-submodule such that $N/C_N(A)$ is FC-hypercentral in G.

Theorem 1

If either

(i) A is RG-noetherian and $H_0(G,A) = 0$

or

(ii) A is RG-artinian and $H^0(G,A) = 0$,

then $H_n(K,A) = 0 = H^n(K,A)$ for all $n \geq 0$ and all subgroups K intermediate between N and G.

The idea of the proof is to find a finitely generated subgroup X_0 such that $H_0(X_0,A) = 0$ (or $H^0(X_0,A) = 0$). Then, using the Lyndon-Hochschild-Serre spectral sequence, one proves that $H^n(X,A)=0$ for all X in $\underline{S}$, the set of finitely generated subgroups of G containing X_0. The result will follow on applying Proposition 2 to $\underline{S}$.

Finite rank theorems

Let R be an integral domain with field of fractions F. The torsion-free rank of an R-module A is defined to be

$$r_0(A) = \dim_F(A \otimes_R F).$$

Using Proposition 2 and methods developed in [17], one can prove

Theorem 2.

Let G be a locally nilpotent group, R a noetherian domain whose non-zero prime ideals are maximal and A an RG-module which is torsion-free with finite rank as an R-module. If $H_0(G,A) = 0$, then $H_n(G,A) = 0 = H^n(G,A)$ for all $n \geq 0$.

Consider next the situation for torsion R-modules. First the concept of rank needs clarification.

Let R be a Dedekind domain and A an R- module which is not torsion-free. Then $\mathrm{Ass}_R(A)$, the associated set of prime ideals, contains a non-zero prime P. Note that P is maximal in R (for this and other facts about Dedekind domains see [2]).

Define the P-primary component of A

$$A_P$$

to be the set of all a in A such that $aP^k = 0$ for some $k > 0$. Then A_P is a submodule of A containing the submodule

$$A[P] = \{a \in A \mid aP = 0\}.$$

Of course A[P] is a vector space over the field R/P. Thus one may define the P-rank of A to be

$$r_P(A) = \dim_{R/P}(A[P]).$$

The total rank of A is

$$r_{tot}(A) = r_0(A) + \sum_P r_P(A),$$

the sum being over all P in $Ass_R(A)$: it can be shown that a torsion module has finite total rank if and only if it is artinian.

Theorem 3

Let G be a locally nilpotent group, R a Dedekind domain and A an RG-module which is torsion with finite total rank as an R-module. If $H^0(G,A) = 0$, then $H_n(G,A) = 0 = H^n(G,A)$ for all $n \geq 0$.

The dual nature of Theorems 2 and 3 is evident. The two theorems may be combined to give a result for mixed modules.

Theorem 4

Let G be a locally nilpotent group, R a Dedekind domain and A an RG-module with finite total rank as an R-module. Then the following are equivalent.

(i) $H^0(G,A) = 0 = H^1(G,A)$,

(ii) $H_0(G,A) = 0 = H^0(G,A)$,

(iii) $H_0(G,A) = 0 = H_1(G,A)$,

(iv) $H_n(G,A) = 0 = H^n(G,A)$ for all $n \geq 0$.

The splitting and conjugacy cases of this theorem generalize [11], Theorem A'.

Modules with infinitely many primary components

If A is a module over a Dedekind domain R, we shall say that A has finite R-ranks if $r_0(A)$ and $r_P(A)$ are finite for all $P \in Ass_R(A)$. This is a weaker property than that of having finite total rank. It seems that no vanishing theorems have been proved for modules of this type. We announce two results of this kind.

Theorem 5

Let G be a countable locally nilpotent group, R a Dedekind domain and A an RG-module which is torsion with finite ranks as an R-module. If $H^0(G,A) = 0$, then either $H^1(G,A)$ is uncountable or else $H^n(G,A) = 0$ for all $n \geq 0$.

In particular, if $H^0(G,A) = 0 = H^1(G,A)$, then $H^n(G,A) = 0$ for all $n \geq 0$. Results like this are frequently useful in the study of complete groups (see [18] in this connection).

It should be emphasized that is possible in the situation of Theorem 5 for $H^n(G,A)$ to be uncountable for all $n \geq 1$ - see below for an example.

Next we mention a result which is applicable to locally nilpotent torsion groups with cardinal a finite aleph. If G is a torsion group, let $\pi(G)$ denote the set of prime divisors of the orders of element of G.

Theorem 6

Let G be a locally nilpotent torsion group with cardinal $\aleph_m$ where m is finite, let R be a Dedekind domain whose characteristic does not belong to $\pi(G)$ and let A be an RG-module. Assume that A is torsion with finite ranks as an R-module. If $H^0(G,A) = 0$, then $H^n(G,A) = 0$ for all $n \geq m+2$.

Taking $R = \mathbb{Z}$ and $m = 0$ we obtain the splitting theorem of Hartley referred to in §2. We shall give a proof of Theorem 6 since it illustrates the power of the spectral sequence for direct limits. In the proof we shall also require two results about the functors $\varprojlim^{(n)}$.

(a) (Goblot [7]). If $\{A_i \mid i \in I\}$ is an inverse system of abelian groups and I has cardinal $\aleph_m$ where m is finite, then $\varprojlim^{(n)}(A_i) = 0$ for all $n \geq m+2$.

(b) (Jensen [15]). If (A_i) is an inverse system of artinian R-modules where R is any ring, then $\varprojlim^{(n)}(A_i) = 0$ for all $n > 0$.

Proof of Theorem 6

Let $\{S_i \mid i \in I\}$ be the set of all finite subgroups of G, ordered by inclusion. Then (S_i) is a direct system of groups, the homomorphisms being inclusions, and $\varinjlim(S_i) = G$. By the spectral sequence it will suffice to prove that $E_2^{pq} = \varprojlim^{(p)}(H^q(S_i,A)) = 0$ provided that $p+q = n \geq m+2$.

Let $q > 0$. Since S_i is finite, $H^q(S_i,A) \simeq \coprod_P H^q(S_i,A_P)$, the sum being over all P in $\mathrm{Ass}_R(A)$. Now $H^q(S_i,A)$ is annihilated by

$|S_i|$ and $|S_i|$ is not 0 in R. It follows that $H^q(S_i,A_P) = 0$ for almost all P in $Ass_R(A)$. Further $H^q(S_i,A_P)$ is isomorphic with a subquotient of a finite direct power of A_P, so it is an artinian R-module. Consequently $H^q(S_i,A)$ is artinian for each i in I. By Jensen's theorem above $E_2^{pq} = 0$ if $p > 0$ and $q > 0$.

Next $E_2^{n0} = \varprojlim^{(n)}(A^{S_i}) = 0$ by Goblot's theorem since $|I| = |G| = \aleph_m$ and $n \geq m+2$.

Finally consider

$$E_2^{0n} = \varprojlim(H^n(S_i,A)) = \varprojlim(\coprod_P H^n(S_i,A_P)).$$

Since $\varprojlim$ is left exact and commutes with products,

$$E_2^{0n} \hookrightarrow \prod_P(\varprojlim(H^n(S_i,A_P))).$$

Now fix P. Since A_P is artinian, there is an $i \in I$ for which $(A_P)^{S_i}$ is minimal. But then $(A_P)^{S_i} = 0$ since $A^G = 0$. Thus $H^n(S_j,A_P) = 0$ for all S_j with $S_i \subseteq S_j$ by Theorem 3. It follows that $\varprojlim(H^n(S_i,A_P)) = 0$ and $E_2^{0n} = 0$, as required.

We mention a recent result of Holt on locally finite groups which can also be proved with the aid of the spectral sequence, but rather more easily.

Theorem 7 ([13])
Let G be a locally finite group with cardinal $\aleph_m$, $m < \infty$, and A a $\mathbb{Z}G$-module which is torsion as a $\mathbb{Z}$-module. Assume that $\pi(A) \cap \pi(G)$ is empty. Then $H^n(G,A) = 0$ for all $n \geq m+2$.

An example

To conclude let us show that in the situation of Theorem 5 it is possible for cohomology to be uncountable in all positive dimensions.

Consider two infinite sequences of distinct primes $p_1, p_2, \ldots$, $q_1, q_2, \ldots$ such that q_i divides $p_i - 1$. Let $\langle a_i \rangle$ and $\langle x_i \rangle$ be cyclic groups of orders p_i and q_i respectively, and define

$$A = \coprod_{i=1,2,..} \langle a_i \rangle \text{ and } X = \coprod_{i=1,2,..} \langle x_i \rangle.$$

The natural action of x_i on a_i and the trivial action of x_i on a_j, $j \neq i$, afford A the structure of a $\mathbb{Z}X$-module. By Theorem 6 $H^n(X,A) = 0$ if $n \neq 1$. On the other hand, it is not difficult to show that

$$H^1(X,A) \simeq \bar{A}/A$$

where $\bar{A} = \prod_{i=1,2,\ldots} \langle a_i \rangle$; this, of course, is uncountable.

Let $G = X \times F$ where F is a free abelian group of countably infinite rank. If F acts trivially on A, then A becomes a $\mathbb{Z}G$-module. A straightforward calculation yields

$$H^n(G,A) \cong \mathrm{Hom}(H_{n-1}(F), \bar{A}/A), \quad n \geq 1 ,$$

which is uncountable.

This example shows that Theorem 6 does not hold if the group G contains elements of infinite order.

By a similar example one may show that the assumption on the characteristic of R in Theorem 6 cannot be omitted.

References

[1] D.W. Barnes, P. Schmid and U.Stammbach, Cohomological characterizations of saturated formations and homomorphs of finite groups, Comment. Math. Helv. 53 (1978), 165-173.

[2] N. Bourbaki, Commutative Algebra, Addison-Wesley, Reading (1972).

[3] K.S. Brown and E. Dror, The Artin-Rees property and homology, Israel J. Math. 22 (1975), 93-109.

[4] P.M. Curran, Fixed-point-free actions on a class of abelian groups, Proc. Amer. Math. Soc. 57 (1976), 189-193.

[5] W. Dwyer, Vanishing homology over nilpotent groups, Proc. Amer. Math. Soc. 49 (1975), 8-12.

[6] W. Gaschütz, Zur Erweiterungstheorie der endlichen Gruppen, J. reine angew. Math. 190 (1952), 93-107.

[7] R. Goblot, Sur les dérivés de certaines limites projectives. Application aux modules, Bull. Sci. Math. 94 (1970), 251-255.

[8] R.L. Griess, Fixed point free action and vanishing cohomology, preprint.

[9] A. Grothendieck, Sur quelques points d'algèbre homologique, Tohoku Math. J. 9 (1957), 119-221.

[10] B. Hartley, Splitting over the locally nilpotent residual for a class of locally finite groups, Quart. J. Math. (2) 27 (1976), 395-400.

[11] B. Hartley and M.J. Tomkinson — Splitting over nilpotent and hypercentral residuals, Math. Proc. Cambridge Philos. Soc. 78 (1975), 215-226.

[12] P.J. Hilton and U. Stammbach — A Course in Homological Algebra, Springer, New York (1970).

[13] D.F. Holt — On the cohomology of locally finite groups, Quart. J. Math. (2) 32 (1981), 165-172.

[14] B. Huppert — Endliche Gruppen I, Springer, Berlin (1979).

[15] C.U. Jensen — Les Foncteurs Dérivés de $\varprojlim$ et leurs Application en Theorie des Modules, Lecture Notes in Mathematics, vol. 254, Springer, Berlin (1970).

[16] D.J.S. Robinson — On the cohomology of soluble groups of finite rank, J. Pure Appl. Algebra 6 (1975), 155-164.

[17] ______ — The vanishing of certain homology and cohomology groups, J. Pure Appl. Algebra 7 (1976), 145-167.

[18] ______ — Recent results on finite complete groups, Algebra, Carbondale 1980, pp. 178-185, Lecture Notes in Math. 848, Springer, Berlin (1981).

[19] ______ — Cohomology of locally nilpotent groups, to appear

[20] J.-L. Roque — Annulation des groupes nilpotents de type fini, C.R. Acad. Sci. Paris. Série A, 284 (1977), 1257-1260.

[21] E. Schenkman — The splitting of certain solvable groups, Proc. Amer. Math. Soc. 6 (1955), 286-290.

[22] D.I. Zaĭcev — On extensions of abelian groups, Akad. Nauk Ukr. SSR, Inst. Mat. (1980), 16-40.

[23] ______ — Soluble extensions of abelian groups, Akad. Nauk Ukr. SSR Inst. Mat. (1981), 14-25.

Untergruppenverbände endlicher auflösbarer Gruppen

Roland Schmidt
Mathematisches Seminar der Universität, Olshausenstr. 40,
D 2300 Kiel 1, Bundesrepublik Deutschland

Einleitung

Im Jahre 1951 bewiesen Suzuki und Zappa (unabhängig voneinander), daß die Klasse S der endlichen auflösbaren Gruppen invariant unter Projektivitäten ist, d.h. daß jede Gruppe, deren Untergruppenverband zu dem einer auflösbaren Gruppe isomorph ist, auflösbar ist. Dieses Resultat legt die Frage nahe, welche Eigenschaften auflösbarer Gruppen bei Projektivitäten erhalten bleiben, oder anders ausgedrückt, welche Teilklassen von S ebenfalls invariant unter Projektivitäten sind. Dazu liefert zunächst einmal die klassische Theorie der auflösbaren Gruppen (Stufe, Rang, Fittinglänge usw.), aber natürlich besonders die nach der Entdeckung der Cartergruppen entwickelte Theorie der Formationen, Schunck- und Fittingklassen Kandidaten in Hülle und Fülle. Wir wollen die folgenden vier Probleme behandeln, einen Überblick über die vorhandenen Resultate und die zu ihrer Gewinnung benutzten Methoden geben und einige neue Ergebnisse beweisen.

(I) Welche Klassen auflösbarer Gruppen sind invariant unter Projektivitäten?

(II) Für welche Klassen K kann man sogar eine verbandstheoretische Charakterisierung finden, also eine Klasse L von Verbänden, so daß eine Gruppe G genau dann in K liegt, wenn ihr Untergruppenverband V(G) zu L gehört? Eine solche Klasse ist dann natürlich invariant unter Projektivitäten.

(III) Hat K vernünftige Vererbungseigenschaften, so kann man in einer beliebigen (auflösbaren) Gruppe G auf geeignete Weise zu K assoziierte Untergruppen U(G,K) bilden (etwa K-Residuum, K-Radikal K-Projektoren, K-Injektoren, usw.). Für welche Klassen K werden diese K-Untergruppen bei Projektivitäten richtig abgebildet, d.h.

gilt $U(G,K)^{\varphi} = U(\overline{G},K)$ für jede Projektivität φ von G auf eine Gruppe $\overline{G}$?

(IV) Für welche Klassen K kann man verbandstheoretische Charakterisierungen solcher K-Untergruppen finden, also eine verbandstheoretische Eigenschaft, so daß U(G,K) für jede auflösbare Gruppe G die einzige Untergruppe von G mit dieser Eigenschaft ist? In diesem Fall wird U(G,K) natürlich bei Projektivitäten richtig abgebildet.

Wir stellen in §1 die Hilfsmittel zum Studium dieser Probleme bereit, behandeln in §2 die Klasse S der auflösbaren Gruppen, studieren in §3 Formationen und untersuchen in den restlichen Paragraphen einige klassische Eigenschaften auflösbarer Gruppen. Da die meisten der dort betrachteten Klassen Formationen sind, folgen dann viele der in diesen Paragraphen besprochenen Resultate aus den allgemeinen Sätzen des §3.

Eigenschaften auflösbarer Gruppen zitieren wir aus den Büchern von Gaschütz [G] oder Huppert [H]. Unsere Bezeichnungen sind die allgemein üblichen (s. etwa [H]), außer daß wir $U \cup V$ für das Erzeugnis zweier Untergruppen U und V der Gruppe G schreiben und für $U \leq V$ das Intervall der zwischen U und V liegenden Untergruppen von G mit [V/U] bezeichnen. Da wir nur endliche Gruppen betrachten, bedeutet "Gruppe" immer endliche Gruppe, und da wir i.allg. nur auflösbare Gruppen untersuchen, ist eine "Formation" immer eine Formation auflösbarer Gruppen.

§1. Methoden

Die meisten interessanten Klassen K auflösbarer Gruppen sind durch arithmetische Bedingungen oder über die Existenz von Normalteilern mit gewissen Eigenschaften definiert, und so gut wie alle Definitionen von K-Untergruppen haben mit Normalteilern zu tun. Deshalb ist es für die Behandlung der erwähnten Probleme entscheidend, die arithmetische Struktur und die Normalteiler von Gruppen verbandstheoretisch in den Griff zu bekommen. Dies gelang Suzuki für die arithmetische Struktur sowie für Normalteiler in nicht auflösbaren Gruppen, während beliebige Normalteiler vor allem von Schmidt untersucht wurden. Wir geben hier nur die Hauptergebnisse an, die wir später benutzen werden.

1.1 Projektivitäten und arithmetische Struktur

Sei p eine Primzahl, $n \in \mathbb{N}$, $n \geq 2$. Die Gruppe G liegt in der Klasse P(n,p), wenn G elementarabelsch der Ordnung p^n oder semidirektes Produkt eines elementarabelschen Normalteilers A der Ordnung p^{n-1} und einer Gruppe $\langle t \rangle$ von Primzahlordnung $q \neq p$ ist, so daß ein $r \in \mathbb{Z}$ existiert mit $t^{-1}at = a^r$ für alle $a \in A$; in diesem Falle ist q ein Teiler von p-1. Die Gruppe G ist eine P-Gruppe, wenn es p und n gibt, so daß $G \in P(n,p)$ ist. Die Gruppen in P(n,p) sind genau die zu der elementarabelschen Gruppe der Ordnung p^n verbandsisomorphen (Baer [1939]).

Die Projektivität φ von G auf $\overline{G}$ heißt
indexerhaltend, wenn $|U^{\varphi}| = |U|$ für alle $U \leq G$ gilt;
singulär, wenn φ nicht indexerhaltend ist;
singulär bei p, wenn es $P \in Syl_p(G)$ mit $|P^{\varphi}| \neq |P|$ gibt;
regulär bei p, wenn φ nicht singulär bei p ist.

Es ist leicht zu sehen, daß es zu einer singulären Projektivität φ immer eine Primzahl p geben muß, für die φ singulär bei p ist. Das Hauptergebnis über solche Projektivitäten (so weit wir es brauchen) lautet:

Satz (Suzuki [1951]). Sei φ eine Projektivität der Gruppe G auf die Gruppe $\overline{G}$ und sei P eine p-Sylowgruppe von G mit $|P^{\varphi}| \neq |P|$. Dann gilt (a) oder (b):

(a) Es existiert eine P-Zerlegung (S,T) von G mit $P < S$, d.h. es ist $G = S \times T$ mit einer P echt enthaltenden P-Gruppe S und $(|S|,|T|) = 1$.

(b) Es gibt ein normales p-Komplement N in G mit G/N zyklisch oder elementarabelsch und $N^{\varphi} \trianglelefteq \overline{G}$.

1.2 Modulare Untergruppen

Der bei der verbandstheoretischen Behandlung der Normalteiler zentrale Begriff ist der der modularen Untergruppe bzw. des modularen Elementes eines Verbandes. Das Element M des Verbandes V heißt modular in V, wenn gilt

(1) $(U \cup M) \cap V = U \cup (M \cap V)$ für alle $U,V \in V$ mit $U \leq V$ und

(2) $(U \cup M) \cap V = (U \cap V) \cup M$ für alle $U,V \in V$ mit $M \leq V$;

die Untergruppe M von G ist modular in G, wenn M modulares Element in $V(G)$ ist. Natürlich sind Normalteiler modulare Untergruppen; die Eigenschaften (1) und (2) sind das, was man vom Normalteiler im Verband erkennen kann. Die grundlegenden Eigenschaften modularer Untergruppen M findet man in Schmidt [1969]; wir benutzen vor allem den Isomorphiesatz

(3) $[U \cup M/M] \simeq [U/U \cap M]$ für alle $U \leq G$.

Um später eine kürzere Sprechweise zur Verfügung zu haben, nennen wir (G,H,K) ein P-hyperzentrales Tripel, wenn H und K Normalteiler der Gruppe G sind und

$$G/K = S_1/K \times \ldots \times S_r/K \times T/K$$

mit $0 \leq r \in \mathbb{Z}$ ist, wobei für $i \neq j \in \{1,\ldots,r\}$ gilt:

(a) S_i/K ist eine P-Gruppe,

(b) $(|S_i/K|,|S_j/K|) = 1 = (|S_i/K|,|T/K|)$ für $i \neq j$,

(c) $H/K = S_1/K \times \ldots \times S_r/K \times (H \cap T)/K$ und

(d) $(H \cap T)/K$ ist hyperzentral in G eingebettet.

Wir erinnern daran, daß ein Faktor X/Y der Gruppe G genau dann hyperzentral (bzw. überauflösbar) in G eingebettet ist, wenn es Normalteiler N_i von G gibt mit

$$Y = N_0 \leq N_1 \leq \ldots \leq N_s = X$$

und $[N_i,G] \leq N_{i-1}$ (bzw. $|N_i : N_{i-1}| \in \mathbb{P}$) für $i=1,\ldots,s$.
Das Hauptresultat über modulare Untergruppen zeigt, daß sie nicht allzuweit vom Normalteiler-Sein entfernt sind.

Satz (Schmidt [1975]). Sei M eine modulare Untergruppe der Gruppe G, sei $H = M^G = \langle M^x | x \in G\rangle$ die normale Hülle und $K = M_G = \bigcap_{x \in G} M^x$ das Herz von M in G. Dann ist (G,H,K) ein P-hyperzentrales Tripel.

Insbesondere folgt:

Korollar. Ist M modular in G, so ist M^G/M_G überauflösbar in G eingebettet.

1.3 Bilder von Normalteilern unter Projektivitäten

Im allgemeinen werden Normalteiler von Projektivitäten zwar nicht auf Normalteiler, aber doch auf modulare Untergruppen abgebildet. Der Satz des letzten Abschnitts liefert daher einen ähnlich aussehenden für Bilder von Normalteilern unter Projektivitäten.

Satz 1 (Schmidt [1975]). Sei N ein Normalteiler der Gruppe G, φ eine Projektivität von G auf die Gruppe $\overline{G}$, $H^{\varphi} = (N^{\varphi})^{\overline{G}}$ die normale Hülle und $K^{\varphi} = (N^{\varphi})_{\overline{G}}$ das Herz von N^{φ} in $\overline{G}$. Dann sind (G,H,K) und $(\overline{G},H^{\varphi},K^{\varphi})$ P-hyperzentrale Tripel.

Speziell für minimale Normalteiler gilt:

Satz 2 (Schmidt [1975]). Sei N ein minimaler Normalteiler der Gruppe G und φ eine Projektivität von G auf die Gruppe $\overline{G}$, so daß N^{φ} kein minimaler Normalteiler von $\overline{G}$ ist. Dann ist N zyklisch von Primzahlordnung p, und mit $S^{\varphi} = (N^{\varphi})^{\overline{G}}$ gilt eine der folgenden beiden Aussagen:

(a) S und S^{φ} sind elementarabelsche p-Gruppen mit $S \leq Z(G)$ und $S^{\varphi} = N^{\varphi} \times (S^{\varphi} \cap Z(\overline{G})) \leq Z_2(\overline{G})$.

(b) S und S^{φ} sind P-Gruppen, und es ist $G = S \times T$ sowie $\overline{G} = S^{\varphi} \times T^{\varphi}$ mit $(|S|,|T|) = 1 = (|S^{\varphi}|,|T^{\varphi}|)$.

Satz 3 (Schmidt [1973]). Sei φ eine Projektivität von G auf $\overline{G}$ und N ein minimaler Normalteiler von G. Ist N zyklisch, etwa $|N| = p$, so sei ferner $N^{\varphi} \trianglelefteq \overline{G}$ und φ sowie φ^{-1} regulär bei p. Dann ist $C_G(N)^{\varphi} = C_{\overline{G}}(N^{\varphi})$.

1.4 Verbandstheoretische Charakterisierungen

Es liegt nahe zu versuchen, verbandstheoretische Charakterisierungen für Gruppenklassen K oder K-Untergruppen dadurch zu gewinnen, daß man die gruppentheoretischen Eigenschaften, die in der Definition der Klasse oder der speziellen Untergruppe auftreten und die nicht bereits verbandstheoretischer Natur sind, durch diejenigen verbandstheoretischen Eigenschaften ersetzt, die sie charakterisieren oder am besten approximieren. So wird man etwa "Normalteiler" durch "modulare Untergruppe", "zyklische Faktorgruppe H/K" durch "distributiven Faktorverband [H/K]" und "abelsche Faktorgruppe H/K" durch "modularen Faktorverband [H/K]" zu ersetzen versuchen. Das wird zwar nicht immer funktionieren - so ist ja bereits eine Gruppe nicht genau dann abelsch, wenn sie modularen Untergruppenverband hat -, liefert aber doch recht häufig nette verbandstheoretische Charakterisierungen. Ein Beispiel für eine solche Charakterisierung, die sofort aus 1.2 folgt, ist etwa: Die Gruppe G ist genau dann einfach, wenn ihr Untergruppenverband außer 1 und G kein modulares Element besitzt. Die meisten unserer verbandstheoretischen Charakterisierungen werden nach diesem Verfahren hergestellt sein.

§ 2. Die auflösbaren Gruppen

Für die Klasse S der auflösbaren Gruppen sind die in der Einleitung betrachteten Probleme seit langem gelöst.

2.1 Satz (Suzuki [1951], Zappa [1951]). Ist G eine auflösbare Gruppe und φ eine Projektivität von G auf eine Gruppe $\bar{G}$, so ist auch $\bar{G}$ auflösbar, d.h. S ist invariant unter Projektivitäten.

2.2 Verbandstheoretische Charakterisierungen

(a) (Suzuki [1956]) G ist genau dann auflösbar, wenn eine Kette $1 = U_o < U_1 < \ldots < U_r = G$ von Untergruppen U_i existiert mit

(i) $U_i^\sigma = U_i$ für jede Autoprojektivität σ von G und

(ii) $[U_{i+1}/U_i]$ modular für alle $i=0,\ldots,r-1$.

Eine Übersetzung der üblichen Definitionen für Auflösbarkeit im Sinne des in 1.4 angedeuteten Verfahrens ist:

(b) (Schmidt [1968]) Die folgenden Eigenschaften der Gruppe G sind äquivalent:

(i) G ist auflösbar.

(ii) Es gibt Untergruppen G_i von G mit $1 = G_o \leq \ldots \leq G_r = G$, G_i modular in G und $[G_{i+1}/G_i]$ modular $(i=0,\ldots,r-1)$.

(iii) Es gibt Untergruppen G_i von G mit $1 = G_o \leq \ldots \leq G_s = G$, G_i modular in G_{i+1} und $[G_{i+1}/G_i]$ modular $(i=0,\ldots,s-1)$.

(iv) Es gibt Untergruppen G_i von G mit $1 = G_o < \ldots < G_t = G$, G_i maximal und modular in G_{i+1} $(i=0,\ldots,t-1)$.

2.3 Radikal und Residuum

In einer beliebigen endlichen Gruppe G definiert man das auflösbare Radikal G_S als das Produkt aller auflösbaren Normalteiler von G und das auflösbare Residuum G^S als Durchschnitt aller Normalteiler von G mit auflösbarer Faktorgruppe. Dann ist offenbar G_S der größte auflösbare Normalteiler und G^S der kleinste Normalteiler mit auflösbarer Faktorgruppe von G.

Satz (Suzuki [1951]). Ist φ eine Projektivität von G auf $\overline{G}$, so ist $(G^S)^\varphi = \overline{G}^S$ und $(G_S)^\varphi = \overline{G}_S$.

2.4 Verbandstheoretische Charakterisierungen

Es ist leicht zu sehen und wohlbekannt, daß das auflösbare Radikal der Durchschnitt aller maximalen auflösbaren Untergruppen und das auflösbare Residuum die maximale perfekte Untergruppe von G ist. Da die auflösbaren Gruppen mit 2.2 charakterisiert sind und die perfekten Gruppen nach 1.2 gerade die Gruppen G sind, deren sämtliche maximalen Untergruppen nicht modular in G sind, erhält man über die Bemerkung sofort verbandstheoretische Charakterisierungen von G_S und G^S. Wir verzichten darauf, sie ausführlich zu formulieren, und geben lieber die folgende, einfachere Charakterisierung an.
Wir nennen einen Verband V polymodular, wenn er Nullelement O und Allelement I besitzt und es Elemente $M_i \in V$ gibt mit $O = M_o \leq \ldots \leq M_r = I$, M_i modular in V und $[M_{i+1}/M_i]$ modular $(i=0,\ldots,r-1)$.

Satz. Sei G eine Gruppe. Dann ist G_S die größte modulare Untergruppe mit polymodularem Untergruppenverband und G^S die kleinste modulare Untergruppe mit polymodularem Faktorintervall $[G/G^S]$ von G.

Beweis. Nach 2.2 ist G^S eine modulare Untergruppe mit polymodularem Faktorintervall $[G/G^S]$. Ist M irgendeine solche Untergruppe, so ist $[G/M^G]$ polymodular und folglich G/M^G nach 2.2 auflösbar. Da auch M^G/M_G auflösbar ist, ist G/M_G auflösbar und somit $M \geq M_G \geq G^S$. Es ist also G^S die kleinste modulare Untergruppe mit modularem Faktorverband. Die Charakterisierung von G_S beweist man genauso.

§ 3. Formationen auflösbarer Gruppen

3.0 Definitionen und Beispiele

Eine Klasse F auflösbarer Gruppen heißt Formation, wenn sie die folgenden Eigenschaften hat:

(a) Ist $N \trianglelefteq G \in F$, so ist $G/N \in F$.

(b) Sind $N_1, N_2 \trianglelefteq G$ mit $G/N_i \in F$ für $i=1,2$, so ist $G/N_1 \cap N_2 \in F$.

Eine Formation F heißt gesättigt, wenn aus $G/\phi(G) \in F$ folgt $G \in F$.

Zu jeder Formation F und jeder auflösbaren Gruppe G existiert wegen (b) ein eindeutig bestimmter kleinster Normalteiler von G, dessen Faktorgruppe in F liegt, das F-Residuum G^F; es ist der Durchschnitt aller Normalteiler N von G mit $G/N \in F$. Ferner existiert zu jeder gesättigten Formation F und jeder auflösbaren Gruppe G eine kanonische Konjugiertenklasse von Untergruppe S von G, sogenannte F-Projektoren, die definiert sind durch die Eigenschaft, daß SN/N F-maximal in G/N ist für jeden Normalteiler N von G ([G], S. 15 und 36). Ein einfaches Konstruktionsverfahren für (gesättigte) Formationen liefert:

Satz (Gaschütz [1963]). Für jede Primzahl p sei $F(p)$ eine Klasse von Gruppen. Dann ist die Klasse F der auflösbaren Gruppen G mit $G/C_G(H/K) \in F(p)$ für jeden p-Hauptfaktor H/K von G eine Formation; wir nennen sie die durch die $F(p)$ lokal definierte Formation. Sind alle $F(p)$ Formationen, so ist F gesättigt.

Beispiele. (1) Die Klasse A der abelschen Gruppen ist offenbar eine Formation. Da es nichtabelsche p-Gruppen gibt, ist A nicht gesättigt.

(2) Die Klasse N der nilpotenten Gruppen bildet eine gesättigte Formation: sie wird lokal definiert durch die Formationen $F(p) = \{1\}$ für alle $p \in \mathbb{P}$.

(3) Sei K eine Formation, $F(p) = K$ für alle $p \in \mathbb{P}$ und sei F die durch diese $F(p)$ lokal definierte Formation. Ist $G \in F$, so ist $G/F(G) \in K$, da die Fittinggruppe $F(G)$ der Durchschnitt der Zentralisatoren der Hauptfaktoren von G ist ([H], S. 278). Somit ist $G \in NK$, der Klasse aller Gruppen mit nilpotentem K-Residuum. Ist

umgekehrt $G \in NK$, so ist $G/F(G) \in K$ und folglich $G \in F$. Damit ist $F = NK$. Insbesondere sind die Klassen NA der Gruppen mit nilpotenter Kommutatorgruppe sowie die Klassen N^k der Gruppen mit Fittinglänge höchstens k gesättigte Formationen.

(4) Für $k \in \mathbb{N}$ sei R_k die Klasse der auflösbaren Gruppen, deren Hauptfaktoren alle höchstens die Dimension k haben. Dann sind offenbar alle R_k Formationen, aber nur die Formation R_1 der überauflösbaren Gruppen ist gesättigt: sie wird lokal definiert durch die Formationen A_p der abelschen Gruppen mit p-1 teilendem Exponenten (s. [H], S. 713 und [G], S. 43).

(5) Für $p \in \mathbb{P}$ sei Z_p die aus der Einsgruppe und sämtlichen zyklischen Gruppen Z_q von Primzahlordnung q mit $q|p-1$ bestehende Klasse. Die durch diese Z_p lokal definierte Formation besteht dann aus denjenigen überauflösbaren Gruppen, die auf jedem Hauptfaktor eine Automorphismengruppe von höchstens Primzahlordnung induzieren; dies ist bekanntlich die Klasse M_L der Gruppen mit nach unten semimodularem Untergruppenverband (Suzuki [1956], S. 10).

(6) Für $p \in \mathbb{P}$ sei Q_p die Formation der abelschen Gruppen mit quadratfreiem p-1 teilendem Exponenten. Die durch die Q_p lokal definierte gesättigte Formation Q liegt offenbar zwischen M_L und R_1.

3.1 Invarianz unter Projektivitäten

Satz. Für jede Primzahl p sei $F(p)$ eine Klasse auflösbarer Gruppen mit den folgenden Eigenschaften:

(1) $Z_p \subseteq F(p)$.

(2) Sind X,Y verbandsisomorphe irreduzible Untergruppen von $GL(n,p)$, $n \in \mathbb{N}$, und ist $X \in F(p)$, so ist auch $Y \in F(p)$.

Dann ist die durch die $F(p)$ lokal definierte Formation F invariant unter Projektivitäten.

Beweis. Sei $G \in F$ und φ eine Projektivität von G auf $\overline{G}$; wir zeigen $\overline{G} \in F$ mittels Induktion nach $|G|$.

Ist G P-zerlegbar, also $G = S \times T$ mit $(|S|,|T|) = 1$ und P-Gruppe S, so ist $\overline{G} = S^\varphi \times T^\varphi$, ferner $T \simeq G/S \in F$ und nach Induktionsannahme also auch $\overline{G}/S^\varphi \simeq T^\varphi \in F$. Ist nun H/K ein p-Hauptfaktor von $\overline{G}$ mit $K \geq S^\varphi$, so ist also $\overline{G}/C_{\overline{G}}(H/K) \in F(p)$; ist $H \leq S^{\varphi}$, so ist $\overline{G}/C_{G}(H/K) \in Z_p \subseteq F(p)$, da S^{φ} eine P-Gruppe ist. Damit ist $\overline{G} \in F$. Wir können also annehmen, daß G nicht P-zerlegbar ist.

Sei nun N ein minimaler Normalteiler von G. Ist N^{φ} nicht normal in $\overline{G}$, so existiert nach 1.3 eine minimale Untergruppe M von Z(G) mit $M^{\varphi} \leq Z(\overline{G})$; sei $|M^{\varphi}| = q$. Nach Induktionsannahme ist $\overline{G}/M^{\varphi} \in F$, ferner $\overline{G}/C_{\overline{G}}(M^{\varphi}) = 1 \in F(q)$ und damit $\overline{G} \in F$.

Sei also $N^{\varphi} \trianglelefteq \overline{G}$ und $|N| = p^n$. Nach 1.3 ist dann N^{φ} ein minimaler Normalteiler von $\overline{G}$, und nach Induktionsannahme ist $\overline{G}/N^{\varphi} \in F$.

Ist $n \geq 2$ oder $n = 1$ und φ sowie φ^{-1} regulär bei p, so ist nach 1.3 ferner $C_G(N)^{\varphi} = C_{\overline{G}}(N^{\varphi})$, und φ induziert eine Projektivität von $X = G/C_G(N)$ auf $Y = \overline{G}/C_{\overline{G}}(N^{\varphi})$. Beide Gruppen sind wegen $|N| = p^n = |N^{\varphi}|$ irreduzible Untergruppen von GL(n,p), wegen $G \in F$ ist $X \in F(p)$, nach Voraussetzung also auch $Y = \overline{G}/C_{\overline{G}}(N^{\varphi}) \in F(p)$. Erneut ist $\overline{G} \in F$.

Es bleibt der Fall zu betrachten, daß $N^{\varphi} \trianglelefteq \overline{G}$, $|N| = p$ und φ oder φ^{-1} singulär bei p ist; sei $|N^{\varphi}| = q$. Ist φ^{-1} singulär bei q, so existiert nach 1.1 ein normales q-Komplement mit abelscher Faktorgruppe in $\overline{G}$, da $\overline{G}$ mit G ebenfalls P-unzerlegbar ist. Dann ist N^{φ} ein zentraler Hauptfaktor, also $\overline{G}/N^{\varphi} = 1 \in F(q)$ und $\overline{G} \in F$. Sei also φ^{-1} regulär bei q. Dann ist $p = q$, φ singulär bei p, und nach 1.1 existiert ein normales p-Komplement K in G mit $K^{\varphi} \trianglelefteq \overline{G}$ und $\overline{G}/K^{\varphi}$ zyklisch oder P-Gruppe. Es folgt $\overline{G}/C_{\overline{G}}(N^{\varphi}) \in Z_p \subseteq F(p)$ und somit wieder $\overline{G} \in F$. Damit ist der Satz bewiesen.

Als Folgerung erhalten wir eine leichte Verallgemeinerung eines Satzes aus (Schmidt [1973]).

Korollar. Für jede Primzahl p sei F(p) eine Klasse auflösbarer Gruppen mit den folgenden Eigenschaften:

(1) $Z_p \subseteq F(p)$.

(2') Ist $X \in F(p) \setminus Z_p$ mit zyklischem Zentrum und Y verbandsisomorph zu X, so ist $Y \in F(p)$.

Dann ist die durch die F(p) lokal definierte Formation F invariant unter Projektivitäten.

Beweis. Wir haben zu zeigen, daß die F(p) die Bedingung (2) erfüllen. Seien also X,Y verbandsisomorphe irreduzible Untergruppen von

GL(n,p) und sei $X \in F(p)$. Dann ist nach dem Schurschen Lemma Z(X) zyklisch. Ist also $X \notin Z_p$, so ist $Y \in F(p)$ nach Voraussetzung (2'). Ist aber $X \in Z_p$, so ist $|X|$ ein Teiler von p-1 und wegen der Irreduzibilität von X folglich $n = 1$ (s. [H], S. 165). Da $Y \leq GL(1,p) \simeq Z_{p-1}$ verbandsisomorph zu X ist, ist also auch Y zyklisch von p-1 teilender Primzahlordnung (oder Y=1) und somit $Y \in Z_p \subseteq F(p)$ nach (1). Damit ist (2) gezeigt und das Korollar bewiesen.

Mit Beispiel (3) aus 3.0 erhält man, daß für eine Formation K von auflösbaren Gruppen, die (1) und (2) oder (1) und (2') an Stelle von F(p) für alle Primzahlen p erfüllt, die Klasse NK der Gruppen mit nilpotentem K-Residuum invariant unter Projektivitäten ist. Das liefert insbesondere die projektive Invarianz der Klassen NA, N^k und N^kA für $k \geq 2$ (s. Schmidt [1973]).

Die Klassen Z_p erfüllen offenbar die Voraussetzungen (1) und (2') des Korollars, und wie im Beweis dieses Korollars zeigt man, daß die Q_p den Voraussetzungen (1) und (2) des Satzes genügen. Die zugehörigen lokal definierten Formationen M_L und Q sind unter den projektiv invarianten ausgezeichnet. Es gilt:

Bemerkung. (a) Ist $F \neq \{1\}$ eine Formation, die lokal durch Klassen F(p) definiert und die invariant unter Projektivitäten ist, so ist $Z_p \subseteq F(p)$ für alle p; die Formation M_L der Gruppen mit nach unten semimodularem Untergruppenverband ist also die kleinste nichttriviale lokal definierbare Formation, die invariant unter Projektivitäten ist.

(b) Die Formation Q ist die kleinste nichttriviale gesättigte Formation, die invariant unter Projektivitäten ist.

(c) Es gilt $M_L \subset Q \subset R_1$.

Beweis. (a) Sei $1 \neq G \in F$. Dann existiert $N \trianglelefteq G$ mit G/N zyklisch von Primzahlordnung. Da F invariant unter Projektivitäten und $G/N \in F$ ist, gilt $Z_p \in F$ für alle $p \in \mathbb{P}$, also $1 \in F(p)$. Mit $Z_p \times Z_p$ liegt für jede p-1 teilende Primzahl q auch die dazu verbandsisomorphe nichtabelsche Gruppe der Ordnung pq in F. Es folgt $Z_q \in F(p)$, d.h. $Z_p \subseteq F(p)$. Damit ist $M_L \subseteq F$, und es gilt (a).

(b) Sei $F \neq \{1\}$ eine gesättigte Formation, die invariant unter Projektivitäten ist. Nach Lubeseder (s. [H], S. 710) existieren Formationen $F(p)$, durch die F lokal definiert wird. Nach (a) ist $Z_p \subseteq F(p)$, und da $F(p)$ eine Formation ist, folgt $Q_p \subseteq F(p)$. Damit ist $Q \subseteq F$, und da Q invariant unter Projektivitäten und gesättigt ist, gilt (b).

(c) Da die lokal definierenden Klassen ineinander enthalten sind, gilt $M_L \subseteq Q \subseteq R_1$. Seien p,q,r Primzahlen mit $q \neq r$ und $q^2 r | p-1$ und seien $G = Z_p Z_{qr}$ sowie $H = Z_p Z_{q^2}$ die semidirekten Produkte der zyklischen Gruppe der Ordnung p mit Untergruppen der Ordnung qr bzw. q^2 ihrer Automorphismengruppe. Dann ist $G \in Q \setminus M_L$ und $H \in R_1 \setminus Q$. Es folgt $M_L \subset Q \subset R_1$.

3.2 Residuen

Um die Residuen der in 3.1 betrachteten Formationen gemeinsam mit denen der Formationen R_k zu behandeln, definieren wir: Die Formation F heiße P-hyperzentral abgeschlossen, wenn für jedes P-hyperzentrale Tripel (G,H,K) aus $G/H \in F$ folgt $G/K \in F$. Beispiele für solche Formationen sind die in 3.1 betrachteten.

Lemma. Ist F eine durch Klassen $F(p)$ lokal definierte Formation und gilt $Z_p \subseteq F(p)$ für alle $p \in \mathbb{P}$, so ist F P-hyperzentral abgeschlossen.

Beweis. Sei (G,H,K) ein P-hyperzentrales Tripel, sei $G/H \in F$ sowie o.B.d.A. $K=1$ und sei X/Y ein p-Hauptfaktor aus einer H enthaltenden Hauptreihe von G. Ist $Y \geq H$, so ist $G/C_G(X/Y) \in F(p)$ wegen $G/H \in F$. Sei also $Y < H$ und dann $X \leq H$. Dann ist X/Y mit den Bezeichnungen aus 1.2 entweder isomorph zu einem in $H \cap T$ gelegenen Hauptfaktor, also zentral in G, d.h. $G/C_G(X/Y) = 1 \in F(p)$, oder X/Y isomorph zu einem in einem S_i gelegenen Hauptfaktor und dann X/Y zentral oder $G/C_G(X/Y) \simeq Z_q$ mit $q \in \mathbb{P}$, $q|p-1$. In jedem Falle ist $G/C_G(X/Y) \in F(p)$, und es folgt $G \in F$.

Satz. Ist F eine Formation, die invariant unter Projektivitäten und P-hyperzentral abgeschlossen ist, so gilt $(G^F)^\varphi = \overline{G}^F$ für jede Projektivität φ einer auflösbaren Gruppe G auf eine Gruppe $\overline{G}$.

Beweis. Sei $G^{F} =: N$. Nach 1.3 ist (G,H,K) mit $H^{\varphi} = (N^{\varphi})^{\overline{G}}$ und $K^{\varphi} = (N^{\varphi})_{\overline{G}}$ ein P-hyperzentrales Tripel. Da $G/N \in F$ ist, liegt auch G/H in F, und da F P-hyperzentral abgeschlossen ist, folgt $G/K \in F$. Da $N = G^{F}$ der kleinste Normalteiler von G mit Faktorgruppe in F ist, muß $K = N$ und somit $N^{\varphi} \trianglelefteq \overline{G}$ sein. Da F invariant unter Projektivitäten ist, liegt $\overline{G}/N^{\varphi}$ in F, und folglich ist $(G^{F})^{\varphi} = N^{\varphi} \geq \overline{G}^{F}$. Wenden wir dieses Resultat auf die auflösbare Gruppe $\overline{G}$ statt G und φ^{-1} statt φ an, so erhalten wir $(\overline{G}^{F})^{\varphi^{-1}} \geq G^{F}$, also die andere Inklusion. Es folgt $(G^{F})^{\varphi} = \overline{G}^{F}$, was zu zeigen war.

Korollar. Sei F eine durch Klassen $F(p)$ lokal definierte Formation mit $Z_p \subseteq F(p)$ für alle p. Ist F invariant unter Projektivitäten (gilt also etwa (2) oder (2') aus 3.1), so ist $(G^{F})^{\varphi} = \overline{G}^{F}$ für jede Projektivität φ einer auflösbaren Gruppe G auf eine Gruppe $\overline{G}$.

Beweis. Nach dem Lemma ist F P-hyperzentral abgeschlossen, und aus dem Satz folgt die Behauptung.

Das vorstehende Korollar verbessert sowohl den Satz als auch das Korollar aus 3.1: bei allen dort betrachteten Formationen werden auch die Residuen durch Projektivitäten richtig abgebildet.

3.3 Projektoren

Für die Projektoren ist die Situation sogar noch besser. Diese existieren nur für gesättigte Formationen (in allen auflösbaren Gruppen), und hier gilt ganz allgemein:

Satz. Sei F eine gesättigte Formation, die invariant unter Projektivitäten ist. Ist S ein F-Projektor der auflösbaren Gruppe G und φ eine Projektivität von G auf eine Gruppe $\overline{G}$, so ist S^{φ} ein F-Projektor von $\overline{G}$.

Beweis. Ist $F = \{1\}$, so ist nichts zu zeigen; sei also $F \neq \{1\}$. Nach Lubeseder (s. [H], S. 710) wird F lokal definiert durch Formationen $F(p)$. Die Bemerkung aus 3.1 liefert $Z_p \subseteq F(p)$ für alle $p \in \mathbb{P}$, nach dem Lemma aus 3.2 ist F P-hyperzentral abgeschlossen.

Sei nun $N^{\varphi} \trianglelefteq \overline{G}$. Wir haben zu zeigen, daß $S^{\varphi}N^{\varphi}/N^{\varphi}$ F-maximal in $\overline{G}/N^{\varphi}$ ist. Da $S \in F$ und F invariant unter Projektivitäten ist, gilt $S^{\varphi} \in F$ und somit $S^{\varphi}N^{\varphi}/N^{\varphi} \simeq S^{\varphi}/S^{\varphi} \cap N^{\varphi} \in F$. Sei $M \leq G$ mit $S^{\varphi}N^{\varphi} \leq M^{\varphi}$ und

M^{φ}/N^{φ} F-maximal in $\overline{G}/N^{\varphi}$. Nach 1.3 gilt für $H = N^G$ und $K = N_G$, daß $(\overline{G},H^{\varphi},K^{\varphi})$ ein P-hyperzentrales Tripel ist. Wegen $M^{\varphi}H^{\varphi}/H^{\varphi} \simeq M^{\varphi}/M^{\varphi} \cap H^{\varphi}$, $N^{\varphi} \leq M^{\varphi} \cap H^{\varphi} \leq M^{\varphi}$ und $M^{\varphi}/N^{\varphi} \in F$ liegt $M^{\varphi}H^{\varphi}/H^{\varphi}$ in F. Ferner ist mit $(\overline{G},H^{\varphi},K^{\varphi})$ auch $(M^{\varphi}H^{\varphi},H^{\varphi},K^{\varphi})$ ein P-hyperzentrales Tripel, und da F P-hyperzentral abgeschlossen ist, folgt $M^{\varphi}H^{\varphi}/K^{\varphi} \in F$. Da F invariant unter Projektivitäten ist, liegt also MH/K in F. Nach Wahl von M ist $S^{\varphi}K^{\varphi} \leq S^{\varphi}N^{\varphi} \leq M^{\varphi}$, also $SK \leq MH$, und da S ein F-Projektor ist, ist SK/K F-maximal in G/K. Es folgt SK=MH und dann $S^{\varphi}N^{\varphi} \geq S^{\varphi}K^{\varphi} \geq M^{\varphi}$, also $S^{\varphi}N^{\varphi} = M^{\varphi}$.

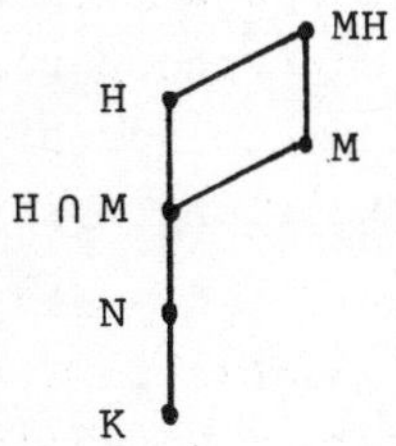

Damit ist $S^{\varphi}N^{\varphi}/N^{\varphi}$ F-maximal in $\overline{G}/N^{\varphi}$ und S^{φ} ein F-Projektor von $\overline{G}$.

3.4 Verbandstheoretische Charakterisierungen

Eine allgemeine verbandstheoretische Charakterisierung der in 3.1 behandelten Formationen ist uns nicht bekannt. Hat man aber für eine einzelne solche Formation F eine verbandstheoretische Charakterisierung - wie etwa für die durch die Klassen Z_p lokal definierte Formation M_L der Gruppen mit nach unten semimodularem Untergruppenverband, so erhält man mit Hilfe des in 1.4 beschriebenen Verfahrens i.allg. auch verbandstheoretische Charakterisierungen für F-Residuum und F-Projektoren.

Satz. Sei F eine Formation auflösbarer Gruppen und L eine Klasse von Verbänden, die F charakterisiert, so daß also eine Gruppe G genau dann in F liegt, wenn $V(G) \in L$ ist. Ist ferner

(1) F P-hyperzentral abgeschlossen und

(2) L abgeschlossen gegen Faktorintervalle nach modularen Elementen, d.h. aus $x m V \in L$ folge $[I/x] = \{y \in V | x \leq y\} \in L$,

so gilt für jede auflösbare Gruppe G:

(a) G^F ist die kleinste modulare Untergruppe von G, deren Faktorverband in L liegt.

(b) Genau dann ist die Untergruppe S von G ein F-Projektor von G, wenn für jede modulare Untergruppe M von G gilt, daß $S \cup M$ L-maximal in [G/M] ist, d.h. $[S \cup M/M]$ in L liegt und $[T/M] \notin L$ ist für alle $S \cup M < T \leq G$.

<u>Beweis</u>. (a) Da G/G^F in F liegt, ist G^F eine modulare Untergruppe von G mit $[G/G^F] \in L$. Ist M irgendeine modulare Untergruppe von G mit $[G/M] \in L$, so ist für $H = M^G$ und $K = M_G$ nach 1.2 (G,H,K) ein P-hyperzentrales Tripel. Nach (2) liegt mit [G/M] auch [G/H] in L, also G/H in F. Nach (1) ist dann auch $G/K \in F$, also $M \geq K \geq G^F$. Damit ist G^F die kleinste modulare Untergruppe von G, deren Faktorverband in L liegt.

(b) Hat S die angegebene Eigenschaft und ist $N \trianglelefteq G$, so ist SN L-maximal in [G/N], also SN/N F-maximal in G/N. Damit ist S ein F-Projektor. Sei umgekehrt S ein F-Projektor und M modular in G.

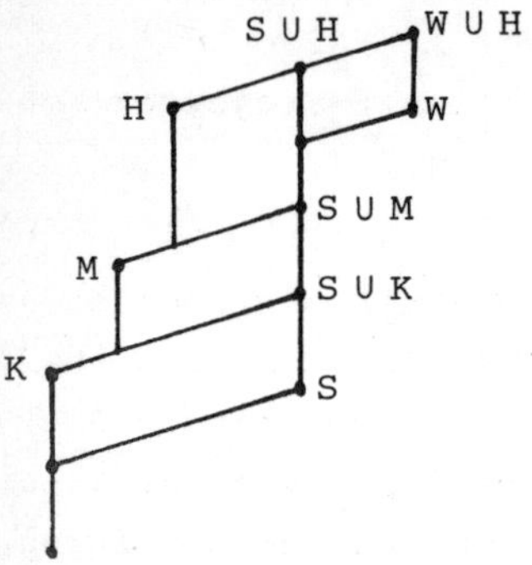

Dann ist wieder (G,H,K) mit $H = M^G$ und $K = M_G$ ein P-hyperzentrales Tripel. Da S ein F-Projektor ist, liegt $S \cup K/K$ in F, also $[S \cup K/K]$ in L. Wegen $S \cup M = (S \cup K) \cup M$ ist $[S \cup M/M] \simeq [(S \cup K)/(S \cup K) \cap M]$, und da $(S \cup K) \cap M$ modular in $[S \cup K/K]$ ist, liegt $[(S \cup K/(S \cup K) \cap M]$ in L, also auch $[S \cup M/M]$ in L. Ist $S \cup M \leq W \leq G$ mit $[W/M] \in L$, so ist $[W \cup H/H] \simeq [W/W \cap H] \in L$ nach (2), also $WH/H \in F$. Da (WH,H,K) ein P-hyperzentrales Tripel und F P-hyperzentral abgeschlossen ist, folgt $WH/K \in F$. Da S ein F-Projektor ist, ist SK/K F-maximal in G/K, wegen $SK \leq WH$ also $SK = WH$ und dann $W \leq WH = SK \leq S \cup M \leq W$, d.h. $S \cup M = W$. Damit ist $S \cup M$ L-maximal in [G/M], was zu zeigen war.

<u>Bemerkung</u>. Das Lemma aus 3.2 zeigt, daß man im vorstehenden Satz (1) ersetzen kann durch

(1') F ist lokal definiert durch Klassen F(p) mit $Z_p \subseteq F(p)$ für alle $p \in \mathbb{P}$.

§ 4. Der Rang einer auflösbaren Gruppe; die Klassen R_k

Sei G eine auflösbare Gruppe und $1 = G_o < \ldots < G_n = G$ eine Hauptreihe von G; sei $|G_i : G_{i-1}| = p_i^{r_i}$ mit $p_i \in \mathbb{P}$ und $r_i \in \mathbb{N}$ für $i = 1,\ldots,n$. Dann nennen wir das ungeordnete n-tupel $(r_1,\ldots,r_n)$

den Rang r(G) von G. Nach dem Satz von Jordan-Hölder ist r(G) unabhängig von der Auswahl der Hauptreihe. Für $k \in \mathbb{N}$ betrachten wir die Klasse R_k der auflösbaren Gruppen G mit $r_i \leq k$ für $i=1,\dots,n$, deren sämtliche Hauptfaktoren also höchstens die Dimension k haben.

<u>4.1 Satz</u> (Schmidt [1972]). Ist φ eine Projektivität der auflösbaren Gruppe $\overline{G}$ auf die (auflösbare) Gruppe $\overline{G}$, so ist $r(\overline{G}) = r(G)$. Insbesondere sind die Klassen R_k für alle $k \in \mathbb{N}$ invariant unter Projektivitäten.

4.2 Verbandstheoretische Charakterisierungen der überauflösbaren Gruppen

Für die Klasse R_1 der überauflösbaren Gruppen sind zwei verschiedene verbandstheoretische Charakterisierungen bekannt. Während die zweite nach dem in 1.4 geschilderten Verfahren gearbeitet ist, stellt die andere einen der seltenen Fälle dar, in dem eine wichtige gruppentheoretische durch eine ebenso interessante verbandstheoretische Eigenschaft gekennzeichnet ist.

(a) (Iwasawa [1941]). Die Gruppe G ist genau dann überauflösbar, wenn ihr Untergruppenverband V(G) die Jordan-Dedekindsche Kettenbedingung erfüllt.

(b) (Schmidt [1960]). Die folgenden Eigenschaften der Gruppe G sind äquivalent:

(i) G ist überauflösbar.

(ii) Es gibt Untergruppen G_i von G mit $1 = G_o \leq \dots \leq G_r = G$, G_i modular in G und $[G_{i+1}/G_i]$ distributiv $(i=0,\dots,r-1)$.

(iii) Es gibt Untergruppen G_i von G mit $1 = G_o \leq \dots \leq G_s = G$, G_i modular in G und maximal in G_{i+1}.

4.3 Verbandstheoretische Charakterisierung von R_k

Die letzte Charakterisierung von R_1 läßt sich verallgemeinern. Wir nennen den Verband V <u>poly-k-modular</u>, wenn er Nullelement O und Allelement I besitzt und Elemente x_i enthält mit $O = x_o \leq \dots \leq x_r = I$, x_i modular in V und $[x_{i+1}/x_i]$ modular der Dimension $\leq k$ $(i=0,\dots,r-1)$. Offenbar ist die Klasse der poly-k-modularen Verbände abgeschlossen gegen Faktorintervalle nach modularen Elementen; denn ist x modular in V, so bilden die $x_i \cup x$ nach (3) aus 1.2 eine Kette von modularen Elementen in $[I/x]$ mit $[x_{i+1} \cup x/x_i \cup x] \simeq [x_{i+1}/x_{i+1} \cap (x_i \cup x)]$, und dieser Verband ist ein Intervall in $[x_{i+1}/x_i]$, also modular der Dimension höchstens k.

Satz. Sei $k \in \mathbb{N}$. Genau dann liegt die Gruppe G in R_k, wenn ihr Untergruppenverband poly-k-modular ist.

Beweis. Ist $G \in R_k$, so bilden die Glieder X_i einer Hauptreihe von G eine Kette mit den gewünschten Eigenschaften. Sei also umgekehrt G eine Gruppe mit poly-k-modularem Untergruppenverband, sei $1 = X_o = \ldots = X_s < X_{s+1} \leq \ldots \leq X_r = G$ mit X_i modular in G und $[X_{i+1}/X_i]$ modular der Dimension $\leq k$, sei $M = X_{s+1}$ und die Behauptung $G \in R_k$ richtig für Gruppen kleinerer Ordnung als $|G|$. Dann ist $V(G/M^G)$ poly-k-modular, nach Induktionsannahme also $G/M^G \in R_k$. Da M modular in G ist, ist M^G/M_G überauflösbar in G eingebettet, und da $V(M_G)$ ein Intervall in $[X_{s+1}/X_s]$, also modular der Dimension $\leq k$ ist, haben alle unterhalb von M_G gelegenen Hauptfaktoren von G ebenfalls höchstens die Dimension k. Damit ist $G \in R_k$.

4.4 Residuen und Projektoren

Ist (G,H,K) ein P-hyperzentrales Tripel mit $G/H \in R_k$, so ist auch $G/K \in R_k$, da die zwischen H und K gelegenen Hauptfaktoren von G alle zyklisch sind. Damit sind alle R_k P-hyperzentral abgeschlossen; die Voraussetzung (2) aus 3.4 ist für die verbandstheoretische Charakterisierung aus 4.3 ebenfalls erfüllt. Mit 3.4 haben wir also:

Satz. Sei $k \in \mathbb{N}$ und G eine auflösbare Gruppe. Dann gilt:

(a) Das R_k-Residuum G^{R_k} ist die kleinste modulare Untergruppe mit poly-k-modularem Faktorverband von G.

(b) Genau dann ist die Untergruppe S ein R_k-Projektor von G, wenn für jede modulare Untergruppe M von G die Gruppe $S \cup M$ maximal in $[G/M]$ mit $[S \cup M/M]$ poly-k-modular ist.

Insbesondere werden R_k-Residuum und R_k-Projektoren (sofern sie existieren) durch Projektivitäten auf R_k-Residuum bzw. R_k-Projektoren abgebildet.

§ 5. Stufe und Fittinglänge

Die Stufe (bzw. Fittinglänge) einer auflösbaren Gruppe G ist die kleinste Zahl k mit $G \in A^k$ (bzw. $G \in N^k$). Daß die Klassen N^k für $k \geq 2$ invariant unter Projektivitäten sind (für k=1 natürlich nicht), ist seit langem bekannt. Es werden sogar die k-ten Fittinggruppen $F_k(G)$, also die Radikale der Fittingklassen N^k, richtig abgebildet.

<u>5.1 Satz</u> (Schmidt [1972]). Ist φ eine Projektivität der Gruppe G auf die Gruppe $\overline{G}$, so ist $F_k(G)^{\varphi} = F_k(\overline{G})$ für $k \geq 2$. Ist also G auflösbar mit Fittinglänge $h(G) \geq 3$, so ist $h(\overline{G}) = h(G)$.

Die Klassen N^k fallen für $k \geq 2$ als Formationen auch unter unsere Sätze des §3. Denn $F(p) = N$ erfüllt (1) und (2') aus 3.1, da nilpotente Gruppen mit zyklischem Zentrum nach Suzuki ([1956], S. 12) nur Projektivitäten auf nilpotente Gruppen zulassen. Damit ist N^2 invariant unter Projektivitäten und mit Induktion folgt dasselbe für N^k mit $k \geq 2$. Genauso erfüllen die $F(p) = A$ die Voraussetzungen aus 3.1 und sind somit die Klassen NA und N^kA invariant unter Projektivitäten. Das Korollar aus 3.2 und der Satz aus 3.3 liefern also:

<u>5.2 Satz</u>. Sei $k \in \mathbb{N}$ mit $k \geq 2$, φ eine Projektivität der auflösbaren Gruppe G auf die Gruppe $\overline{G}$ und sei F eine der Formationen N^k, NA oder N^kA.

(a) Dann ist $(G^F)^{\varphi} = \overline{G}^F$.

(b) Ist S ein F-Projektor von G, so ist S^{φ} ein F-Projektor von $\overline{G}$.

Verbandstheoretische Charakterisierungen der in 5.2 betrachteten Klassen sind nicht bekannt; es scheint uns bereits ein interessantes Problem zu sein, N^2 oder NA zu charakterisieren.

Daß die Stufe k(G) einer auflösbaren Gruppe G bei Projektivitäten i.allg. nicht erhalten bleibt, zeigen einfache Beispiele von Projektivitäten zwischen abelschen und nichtabelschen Gruppen. Aber es sind keine Beispiele bekannt, die zeigen würden, daß sämtliche Klassen A^k nicht invariant unter Projektivitäten sind. Es könnte hier also dieselbe Situation vorliegen wie für die Fittinglänge, daß für $k(G) \geq k$ mit einer festen Zahl k gilt $k(\overline{G}) = k(G)$ für jede zu G verbandsisomorphe Gruppe $\overline{G}$. Bekannt sind bisher nur Abschätzungen für die Stufe von $\overline{G}$; die beste ist:

5.3 Satz (Busetto-Menegazzo [1985]). Sind G und $\overline{G}$ auflösbare Gruppen mit isomorphen Untergruppenverbänden, so ist $k(\overline{G}) \leq 6\, k(G)-4$.

§ 6. Sylowturmgruppen

Zu einer Wohlordnung $<$ der Menge $\mathbb{P}$ der Primzahlen definiert man die Klasse $T_<$ der Gruppen mit einem Sylowturm bzgl. $<$ durch $G \in T_<$ genau dann, wenn es Normalteiler G_i von G gibt mit $1 = G_o \leq \ldots \leq G_r = G$, G_i/G_{i-1} isomorph zu einer p_i-Sylowgruppe von G und $p_r < p_{r-1} < \ldots < p_1$. Alle diese Klassen $T_<$ sind gesättigte Formationen (s. [G], S. 41) und damit auch ihre Vereinigung, die Klasse T der Sylowturmgruppen.

6.1 Satz (Schmidt [1972]). Die Klasse T aller Sylowturmgruppen ist invariant unter Projektivitäten.

Für die einzelnen Klassen $T_<$ gilt das i.allg. nicht; wir zeigen:

6.2 Satz. Genau dann ist die Klasse $T_<$ der Gruppen mit Sylowturm bzgl. $<$ invariant unter Projektivitäten, wenn gilt:

(1) Für alle $p,q \in \mathbb{P}$ mit $q \mid p-1$ ist $q < p$.

Insbesondere ist $T_<$ invariant, wenn $<$ die natürliche Anordnung der Primzahlen ist.

Beweis. Ist (1) nicht erfüllt, so existieren $p,q \in \mathbb{P}$ mit $q \mid p-1$ und $p < q$. Dann liegt die elementarabelsche Gruppe der Ordnung p^2 in $T_<$, aber die dazu verbandsisomorphe P-Gruppe der Ordnung pq nicht.

Sei also die Bedingung (1) erfüllt, sei $G \in T_<$, φ eine Projektivität von G auf $\overline{G}$ und sei die Behauptung $\overline{G} \in T_<$ richtig für Gruppen kleinerer Ordnung als $|G|$. Ist dann G zerlegbar in ein echtes direktes Produkt von Gruppen mit teilerfremden Ordnungen, also $G = H \times K$ mit $H \neq 1 \neq K$ und $(|H|,|K|) = 1$, so ist bekanntlich (s. Suzuki [1956], S. 5) $\overline{G} = H^{\varphi} \times K^{\varphi}$, nach Induktionsannahme $\overline{G}/H^{\varphi} \in T_<$ und $\overline{G}/K^{\varphi} \in T_<$ und wegen der Formationseigenschaft auch $\overline{G}/(H^{\varphi} \cap K^{\varphi}) = \overline{G} \in T_<$. Wir können also annehmen, daß gilt:

(2) Ist $G = H \times K$ mit $(|H|,|K|) = 1$, so ist $H = 1$ oder $K = 1$. Sei p der bzgl. < maximale Primteiler von $|\bar{G}|$, P die p-Sylowgruppe von G, q der bzgl. < maximale Primteiler von $|\bar{G}|$ und Q die q-Sylowgruppe von $\bar{G}$. Ist φ singulär bei p, so ist wegen 1.1 und (2) entweder G eine P-Gruppe, oder es existiert ein normales p-Komplement N in G mit zyklischer oder elementarabelscher Faktorgruppe. Wegen $P \trianglelefteq G$ folgt im letzteren Falle $G = P \times N$ und mit (2) dann $N = 1$. In jedem Falle ist also G und damit auch $\bar{G}$ zyklisch oder eine P-Gruppe, wegen (1) also $\bar{G} \in T_<$. Ist φ^{-1} singulär bei q, so ist, da (2) auch für $\bar{G}$ gilt, genauso $\bar{G}$ zyklisch oder eine P-Gruppe, also $\bar{G} \in T_<$. Wir können daher annehmen, daß φ regulär bei p und φ^{-1} regulär bei q ist. Dann ist p ein Teiler von $|\bar{G}|$, also $p \leq q$ und q ein Teiler von $|G|$, also $q \leq p$. Damit ist $q = p$ und P^{φ} die normale p-Sylowgruppe von $\bar{G}$. Nach Induktionsannahme ist $\bar{G}/P^{\varphi} \in T_<$ und da p der bzgl. < maximale Primteiler von $|\bar{G}|$ ist, folgt $\bar{G} \in T_<$, was zu zeigen war.

Sind die Primzahlen der Wohlordnung < nach durchnumeriert und ist $p = p_n$, so setzen wir $T_<(p) = S_{p_{n-1}} S_{p_{n-2}} \dots S_{p_1}$, wobei S_{p_i} die Klasse der p_i-Gruppen bezeichne. Offenbar wird $T_<$ durch $T_<(p)$ lokal definiert, und es ist $Z_p \subseteq T_<(p)$, wenn (1) gilt. Das Korollar aus 3.2 und der Satz aus 3.3 liefern also auch hier:

<u>6.3 Satz</u>. Sei F eine der Formationen T oder $T_<$ zu einer (1) erfüllenden Wohlordnung < von $\mathbb{P}$ und sei φ eine Projektivität der auflösbaren Gruppe G auf die Gruppe $\bar{G}$.

(a) Dann ist $(G^F)^{\varphi} = \bar{G}^F$.

(b) Ist S ein F-Projektor von G, so ist S^{φ} ein F-Projektor von $\bar{G}$.

Literatur

[G] Gaschütz, W.: Lectures on subgroups of Sylow type in finite soluble groups, Notes on Pure Math. 11, Canberra 1979.

[H] Huppert, B.: Endliche Gruppen I, Springer Verlag, Berlin Heidelberg New York 1967.

Baer, R.:

[1939] The significance of the system of subgroups for the structure of a group, Amer. J. Math. 61, 1-44.

Busetto, G. und Menegazzo, F.:

[1985] Groups with soluble factor groups and projectivities, Rend. Sem. Math. Univ. Padova 73, 249-260.

Gaschütz, W.:

[1963] Zur Theorie der endlichen auflösbaren Gruppen, Math. Z. 80, 300-305.

Iwasawa, K.:

[1941] Über die endlichen Gruppen und die Verbände ihrer Untergruppen, J. Univ. Tokyo 4, 171-199.

Schmidt, R.:

[1968] Eine verbandstheoretische Charakterisierung der auflösbaren und der überauflösbaren endlichen Gruppen, Archiv Math. 19, 449-452.

[1969] Modulare Untergruppen endlicher Gruppen, Illinois J. Math. 13, 358-377.

[1972] Verbandsisomorphismen endlicher auflösbarer Gruppen, Archiv Math. 23, 449-458.

[1973] Lattice isomorphisms and saturated formations of finite soluble groups, Proc. of the Second International Conference on the Theory of Groups (Australian Nat. Univ., Canberra), Lecture Notes in Math., Band 372, 605-610.

[1975] Normal subgroups and lattice isomorphisms of finite groups, Proc. London Math. Soc. (3)30, 287-300.

Suzuki, M:

[1951] On the lattice of subgroups of finite groups, Trans. Amer. Math. Soc. 70, 345-371.

[1956] Structure of a group and the structure of its lattice of subgroups, Springer-Verlag, Berlin Heidelberg New York 1956.

Zappa, G.:

[1951] Sulla risolubilità dei gruppi finiti in isomorfismo reticolare con un gruppo risolubile, Giorn. Mat. Battaglini (4)80, 213-225.

AN EXAMPLE OF A NONABELIAN FROBENIUS-WIELANDT COMPLEMENT

Carlo Maria Scoppola

Dipartimento di Matematica

Università di Trento

38050 Povo, Trento ITALY

1. Let G be a finite group acting on a complex module M. Let N(G,M) be the (normal) subgroup of G generated by all the elements of G that fix some nontrivial vector in M. The factor group G/N(G,M) is called in [LP] a *generalized Frobenius complement* for G; each factor of G/N(G,M) is also called, in [E1], a *Frobenius-Wielandt complement*. The reader is referred to [S2],[E1], for a wider introduction to the topic.

The goal of this paper is to give an example of a p-group G of odd order that has an extra-special Frobenius-Wielandt complement of order p^3 and exponent p. Such an example has been announced in [S2], and is contained in the author's doctoral dissertation [S1]. Here, we construct our example only for p = 3; the advantage in doing so is that the lines of the construction appear much more neatly if one avoids the technicalities that arise in dealing with larger primes.

It should be noted that recently Espuelas has found in [E2] an extra-special Frobenius-Wielandt complement of odd order p^3 and exponent p^2, using a different approach.

We refer here to some well known techniques: in particular, for the notion of "basic commutators" and "Hall commutator collecting process" see [Ha, §12.3]. For any group G, G_i denotes the i-th term of the lower central series of G. Furthermore, the Jennings-Lazard-Zassenhaus p-dimension subgroups of a group G, for a fixed prime p, denoted by $K_i(G)$, play a role here.They are defined by

$$K_i(G) = \prod (G_j)^{p^k}$$

where the product is taken over all indices k, j such that $jp^k \geq i$. An account of their elementary properties can be found on [HB,VIII.1] or on [S1]; a corollary of the following result, which belongs to the "folklore" of the topic, will be useful (see [S1] for a proof):

1.1 *Lemma*. Let p be a prime, F be a free group on the free generators $x_1,\ldots,x_d$. Then all the factor groups $K_i(F)/K_{i+1}(F)$ are elementary abelian, and have a basis whose representatives are:

a) the basic commutators of weight i in the generators, if p does not divide i;

b) the basic commutators of weight i in the generators, and the p-th powers of the representatives of any basis of $K_{i/p}(F)/K_{i/p+1}(F)$, if p divides i.

1.2 *Corollary*. Set p = 3, let F be free on the two generators x and y, and let $G = F/K_{10}(F)$, $H = K_6(F)/K_{10}(F) = K_6(G)$.Then H is elementary abelian, and has a basis whose representatives are the basic commutators in x, y of weights ranging from 6 to 9, and the elements $[y,x]^3$, x^9, y^9, $(xy)^9$ and $(xy^{-1})^9$.

Proof. H is elementary abelian, by [HB, VIII.1.13 (a),(b)], since $K_{10}(G) = 1$. Furthermore, by 1.1, $[y,x,x]$, $[y,x,y]$, x^3, y^3 are representatives of a basis of $K_3(G)/K_4(G)$. By the Zassenhaus identity [H, III.9.7], x^3, y^3, $(xy)^3$ and $(xy^{-1})^3$ are also representatives of a basis of $K_3(G)/K_4(G)$.Another application of 1.1 concludes the proof.///

2. The following lemma gives a purely group-theoretical necessary and sufficient condition for a factor G/N of G to be a Frobenius-Wielandt complement for G.

2.1 *Lemma* . Let G be a p-group, and let N be a normal subgroup of G. Then there is a **C**G-module V such that $N(G,V) \leq N$ if and only if G has a cyclic section H/K such that for all $x \in G - N$ there is a power of x in H - K.

Proof : Suppose first that G has a module V as required.We may assume that V is irreducible: in fact, if M is an irreducible component of V, $N(G,M) \leq N(G,V)$. Then, since G is a p-group,$V = L^G$, where L is a linear module for some subgroup H of G, and let K be the kernel of the action of H on L. Suppose $x \in G - N$, and say μ the character afforded by V, λ the character afforded by L. We have:

$$0 = (\mu_{<x>}, 1_{<x>})$$

Now, by Mackey's theorem and Frobenius reciprocity,

$$(\mu_{<x>}, 1_{<x>}) = (\lambda^G{}_{<x>}, 1_{<x>}) = (\Sigma_y \ (\lambda^y{}_{H^y \cap <x>})^{<x>}, 1_{<x>}) =$$
$$= \Sigma_y \ ((\lambda^y{}_{H^y \cap <x>})^{<x>}, 1_{<x>}) = \Sigma_y \ (\lambda^y{}_{H^y \cap <x>}, 1_{H^y \cap <x>}),$$

where y runs over a set of double cosets representatives of H and <x>. In particular,

$$(\lambda_{H \cap <x>}, 1_{H \cap <x>}) = 1,$$

and therefore $H \cap <x>$ is not contained in K. On the other hand, assume H/K is a section as required. Induce to G a linear character λ of H with kernel K. Set $\mu = \lambda^G$. By the normality of N, $H^g \cap \langle x\rangle$ contains properly $K^g \cap \langle x\rangle$, for all $g \in G$; reading backwards the equalities above, we get the result.///

2.2 *Example.* Set p = 3, let F be free on the two generators x and y, and let $G = F/K_{10}(F)$, $N = K_3(F)/K_{10}(F) = K_3(G)$, $H = K_6(F)/K_{10}(F) = K_6(G)$. We are going to construct a normal subgroup K of H such that H/K is cyclic and every element of G-N has a power in H-K; by 2.1, G/N, which is extra-special of exponent p and order p^3, is isomorphic to a Frobenius-Wielandt complement for G.

We know, from 1.2, that H is elementary abelian. Furthermore, let $g \in G - K_2(G)$. Then g = uv, where u is a product of a power of x by a power of y, and $v \in K_2(G)$. We use the commutator collecting process to compute g^9 and we get $g^9 = (uv)^9 = u^9$, since all the other factors of the expansion disappear modulo $K_{10}(G) = 1$. Therefore g^9 is nontrivial, and lies in one of the subgroups $\langle x^9\rangle$, $\langle y^9\rangle$, $\langle (xy)^9\rangle$ and $\langle (xy^{-1})^9\rangle$ of H.

Similarly, if we pick $g \in K_2(G) - K_3(G)$, we have g = uv, where u is a power of [x,y], and $v \in K_3(G)$. We use the commutator collecting process to get $g^3 = (uv)^3 = u^3$, modulo $K_7(G)$. Thus g^3 is a non trivial element of H, since it is nontrivial modulo $K_7(G)$, by 1.1.

From the above remarks and 1.2 it follows that G_8 does not contain any powers of elements of $G - K_3(G)$, and that G_8 is contained in H. Therefore it is enough to show that there is a normal subgroup K/G_8 of H/G_8 such that H/K is cyclic, and every element of $G/G_8 - K_3(G/G_8)$ has a power in $H/G_8 - K/G_8$; then K will be the required subgroup of G.

To avoid heavy notation, and without any possibility of confusion, from now on we will write G, N, H respectively for G/G_8,

H/G_8, K/G_8 and will indicate the cosets of G_8 simply by their representatives. Note that, with the new notation, every element of $G - K_3(G)$ still has a nontrivial power in H, and $G_8 = K_{10}(G) = 1$.

We now list a few facts about G. They can be proved by straigtforward computation, with the help of the elementary properties of commutators [H, III.1.2] and [HB, VIII.1.13]. For some general results on commutators and powers, that are independent of our restriction on p, and yield the following as a corollary, see [S1].

F1) If $a \in K_2(G)$, $b \in G$, then

$$[a,b^3] = [a,b,b,b][a,b]^3[a,b,b;a,b].$$

F2) If $a,a' \in K_2(G)$, $b,c \in K_3(G)$, then

$$[bc,a,a'] = [b,a,a'][c,a,a'].$$

F3) If a, b are as in F2, then $[b,a^r,a^r] = [b,a,a]^{r^2}$ for every integer r.

By F1, via commutator collectig process, we also get

F4) If b, c are as in F2, we have $(bc)^3 = b^3c^3$.

We now compute g^3, for g in $K_2(G) - K_3(G)$, via commutator collecting process. As above, we have $g = uv$, where u is a power of $[y,x]$, and $v \in K_3(G)$. Then $g^3 = u^3 v^3 [v,u,u][v,u,v]^{-1}$, since all the other factors of the expansion are in $K_{10}(G) = 1$. There is no loss of generality in assuming that v is a product of third powers of elements of G: for, assume $v = v'v''$, where v' is such a product, and $v'' \in K_4(G)$. Via commutator collection, we have $v'^3 = v^3$, and also, by F2, $[v,u,u] = [v',u,u]$, $[v,u,v] = [v',u,v']$. Applying F1 twice, we see that $[v,u,v] = 1$, since $G_8 = 1$. Finally, by F3, we may assume that

$$g^3 = u^3 v^3 [v,[y,x],[y,x]]. \qquad (*)$$

We now set $V = \langle [v,[y,x],[y,x]] \mid v \in K_3(G) \rangle$, and define

$$f\colon K_3(G)/K_4(G) \longrightarrow V$$

$$vK_4(G) \longrightarrow [v,[y,x],[y,x]].$$

We have noted above that f is well defined, and F2 implies that f is a homomorphism. Furthermore, f is injective: for, consider the elements x^3, y^3, $[y,x,x]$, $[y,x,y]$, that are representatives of a basis of $K_3(G)/K_4(G)$, by 1.1. Their images under f are, by F2: $[y,x,x,x,x;y,x]^{-1}$, $[y,x,y,y,y;y,x]^{-1}$, $[y,x,x;y,x;y,x]$, and $[y,x,y;y,x;y,x]$, which are powers of different basic commutators of weight 7, and thus they are independent. Hence, we have that $f(vK_4(G)) = 1$ if and only if $v \in K_4(G)$. Now set

$$V' = \langle v^3\ [v,[y,x],[y,x]] \mid v \in K_3(G)\rangle,$$

and note that V' is contained in the center of G, and therefore is normal in G. V' does not contain the third power of any element of $K_2(G) - K_3(G)$, because V' is contained in $K_7(G)$, and we have seen that those powers are nontrivial modulo $K_7(G)$. On the other hand, assume that $g^9 \neq 1$, $g^9 \in V'$. Since f is a homomorphism, and by F4, $g^9 = v^3[v,[y,x],[y,x]]$, for some $v \in K_3(G)$. Thus $[v,[y,x],[y,x]] \in$ $\in K_9(G)$, and this can happen only if $v \in K_4(G)$, as we have seen above; but then we would easily have $g^9 = 1$, contradiction. Hence V' does not contain any nontrivial 9-th powers.

Again, from now on, we will write G, N, H respectively for G/V', H/V', K/V' and will indicate the cosets of V' simply by their representatives. Note that, with the new notation, every element of $G - K_3(G)$ still has a nontrivial power in H. Again, if we exhibit a maximal subgroup K of H such that all the elements in $G - K_3(G)$ have a power in H - K we are done. But now note that, by (*), every element of $G - K_3(G)$ has a nontrivial power in one of the following cyclic subgroups of H :

$$\langle x^9\rangle,\ \langle y^9\rangle,\ \langle (xy)^9\rangle,\ \langle (xy^{-1})^9\rangle, \langle [y,x]^3\rangle,$$

and H is the direct sum of these five subgroups and of a suitable subgroup K'. Now the subgroup we are looking for is

$$K = \langle K', ab^{-1} \mid a,b \in \{x^9, y^9, (xy)^9, (xy^{-1})^9, [y,x]^3\}\rangle.$$

ACNOWLEDGEMENT

I thank my advisor Prof. George Glauberman for his help, patience and encouragement during the preparation of my thesis, part of which appears in this paper.

REFERENCES

E1 A. Espuelas, *The complement of a Frobenius-Wielandt group*, Proc. London Math. Soc. (3) 48 (1984), 564-576.

E2 A. Espuelas, *A nonabelian Frobenius-Wielandt complement*, to appear.

H B. Huppert, "Endliche Gruppen I", Berlin 1967.

Ha M. Hall, "The theory of groups", II ed., New York 1976.

HB B. Huppert, N. Blackburn, "Finite groups II", Berlin 1982.

LP B. Lou, D. Passman, *Generalized Frobenius Complements*, Proc. Amer. Math. Soc. 17 (1966), 1166-1172.

S1 C.M. Scoppola, Thesis, The University of Chicago.

S2 C.M. Scoppola, *On generalized Frobenius complements*, Proc. Internat.Conf. "Groups - St. Andrews 85", to appear.

SUBNORMAL SUBGROUPS OF FACTORISED GROUPS

Stewart E. Stonehewer
University of Warwick
Coventry CV4 7AL, England

1. INTRODUCTION AND STATEMENT OF MAIN RESULTS

It is well-known that while for a subgroup X of a group G there is always a unique largest subgroup of G containing X as a normal subgroup, namely the normaliser $N_G(X)$, the same is not true in general when normal is replaced by subnormal. Thus when G is the symmetric group of degree 5 there are Sylow 2-subgroups H, K of G with $G=\langle H,K\rangle$ and $X=H\cap K$ of order 2. For example we can take

$$H=\langle(12)(34),\ (13)(24),\ (14)(23),\ (12)\rangle,$$
$$K=\langle(23)(45),\ (24)(35),\ (25)(34),\ (25)\rangle.$$

Then $X=\langle(34)\rangle$ is subnormal in H and in K, but X is not subnormal in G.

Again when G is the alternating group of degree 9 and

$$H=\langle(147),\ (259)\rangle,\quad K=\langle(147),\ (123)(456)(789)\rangle,$$

then $G=\langle H,K\rangle$, H is abelian and K is the wreath product of two groups of order 3. Thus $X=\langle(147)\rangle$ is subnormal in H and in K, but X is not subnormal in G. (See [8].) There are even examples of finite soluble groups G generated by subgroups H and K with H abelian and K dihedral of order 8, X a subgroup of order 2 contained in $H\cap K$ and X not subnormal in G. (See [9].)

On the other hand there is a situation, at least for finite groups, where a common subnormal subgroup of two subgroups is subnormal in their join.

THEOREM 1 (Wielandt [9]). Suppose that a finite group G=HK where H and K are subgroups and let X be a subgroup of $H\cap K$ with X subnormal in H and X subnormal in K. Then X is subnormal in G.

This result was proved first in the case when X is soluble by Maier [3]. When G is infinite it seems unlikely that the Theorem will still be true, but an example illustrating this appears to be difficult to find. Thus when G is a finite p-group, then Theorem 1 is trivial and Wielandt makes use of this fact in his argument. But when G is an infinite p-group, then already there are difficulties in deciding whether or not X is subnormal in G.

The purpose of this work is to indicate for which classes of infinite groups Theorem 1 continues to hold, thereby suggesting where to look for a counterexample. Trivially Theorem 1 holds for nilpotent groups G and therefore soluble groups are an obvious class for investigation. It turns out that there is no counterexample among the metabelian groups. (See Corollary 1 of Lemma 2 in section 2.)

Observe that in two of the above examples X actually lies in the centre of H (because H is abelian). But when G=HK it is easy to see that this phenomenon is excluded. For, suppose that $X \lhd H$ and X is subnormal in K. Then X is subnormal in

$$X^K = X^{HK} = X^G \lhd G$$

and so X is subnormal in G.

When it is known that Theorem 1 holds for <u>certain</u> classes of groups G, then there is a general result which makes it possible to extend the Theorem to <u>certain</u> larger classes. In order to give a precise statement of this result, we recall that the (<u>subnormal</u>) <u>defect</u> of a subnormal subgroup X in a group G is the shortest length d of a series of subgroups

$$X = X_0 \lhd X_1 \lhd \ldots \lhd X_d = G.$$

Then we write $X \lhd^m G$ for any $m \geq d$. For any class $\underline{\underline{X}}$ of groups, denote by $s\underline{\underline{X}}$ the class of subgroups of $\underline{\underline{X}}$-groups and by $Q\underline{\underline{X}}$ the class of quotients of $\underline{\underline{X}}$-groups. If $\underline{\underline{Y}}$ is a second class, then $\underline{\underline{X}}\underline{\underline{Y}}$ is the class of groups G with a normal subgroup N in $\underline{\underline{X}}$ and G/N in $\underline{\underline{Y}}$. We write $\underline{\underline{A}}$ for the class of abelian groups, $\underline{\underline{N}}$ for the class of nilpotent groups and $\underline{\underline{N}}_c$ for the nilpotent groups of class $\leq c$. Then we have

THEOREM 2 (Stonehewer [5]). Let $\underline{\underline{X}} = s\underline{\underline{X}}$ and $\underline{\underline{Y}} = s\underline{\underline{Y}} = Q\underline{\underline{Y}}$ be classes of groups and suppose that $G = HK \in \underline{\underline{A}}\underline{\underline{X}} \cap \underline{\underline{Y}}$ (H and K subgroups) and X a subnormal subgroup of both H and K always implies that X is subnormal in G. Then $G = HK \in \underline{\underline{N}}\underline{\underline{X}} \cap \underline{\underline{Y}}$ and X subnormal in H and K always implies that X is subnormal in G. Moreover suppose that there is an integer $f_1 = f_1(\underline{\underline{X}}, m)$

such that $X \lhd^{f_1} G$ whenever $G=HK \in \underline{\underline{A}}\,\underline{\underline{X}} \cap \underline{\underline{Y}}$ with $X \lhd^m H$ and $X \lhd^m K$. Then there is an integer $f_2=f_2(\underline{\underline{X}}, m, c)$ such that $X \lhd^{f_2} G$ whenever $G=HK \in \underline{\underline{N}}_c \underline{\underline{X}} \cap \underline{\underline{Y}}$ with $X \lhd^m H$ and $X \lhd^m K$.

By taking $\underline{\underline{X}}=\underline{\underline{A}}$ and $\underline{\underline{Y}}$ to be the class of all groups, Theorem 2 and Corollary 1 of Lemma 2 in section 2 give

THEOREM 3. If $G=HK \in \underline{\underline{N}}\,\underline{\underline{A}}$ and if X is subnormal in H and in K, then X is subnormal in G.

Other hypotheses on G sufficient for Theorem 1 to hold can be found by means of Lemma 2 in section 2 in conjunction with Theorem 2 above. We list these in

THEOREM 4 (Stonehewer [5]). Let G=HK with X subnormal in H and in K. Then X is subnormal in G provided

(i) G is nilpotent-by-polycyclic-by-finite, or

(ii) G satisfies the minimal condition for subgroups modulo a nilpotent normal subgroup, or

(iii) G is a soluble minimax group.

In general it seems to be easier, when trying to establish Theorem 1 for an infinite group G, to proceed from a factor group G/N for which the Theorem is known to hold and with N suitably restricted, rather than from a subgroup which is large in some sense in relation to G. Of course one reason for this is the fact that G/N inherits a factorisation from G whereas subgroups do not in general. However, it has been possible to adopt the latter approach with <u>some</u> success and we shall describe this work throughout the remainder of the paper. Denote by $\underline{\underline{F}}$ the class of finite groups. Our objective then is to prove

THEOREM 5. Let G=HK be a periodic group in the class $\underline{\underline{N}}\,\underline{\underline{A}}\,\underline{\underline{F}}$ and let X be a subnormal subgroup of H and of K. Then X is subnormal in G. Moreover let $B \lhd G$ with $B \in \underline{\underline{N}}_c \underline{\underline{A}}$ and G/B finite of order $\leqslant n$ and let $X \lhd^m H$, $X \lhd^m K$. Then there is an integer $f=f(c,m,n)$ such that $X \lhd^f G$.

The restriction that G should be periodic is particularly tiresome and we shall have more to say about this later. However, as an immediate corollary we see that Wielandt's Theorem 1 holds for periodic soluble linear groups G.

Suppose that Theorem 5 has been proved in the case when c=1, i.e. whenever G is a metabelian group extended by a finite group of order $\leq n$. In the notation of Theorem 2 take $\underline{\underline{X}}$ to be the class of groups which are finite of order $\leq n$ modulo a normal abelian subgroup and take $\underline{\underline{Y}}$ to be the class of periodic groups. Then $f(1,m,n)$ from Theorem 5 can be taken as the integer $f_1(\underline{\underline{X}},m)$ in Theorem 2 and this Theorem guarantees the existence of an integer $f_2(\underline{\underline{X}},m,c)$ such that $X \lhd^{f_2} G$ whenever $G=HK \in \underline{\underline{N}}_c\underline{\underline{X}} \cap \underline{\underline{Y}}$ with $X \lhd^m H$ and $X \lhd^m K$. Therefore we can take $f(c,m,n)=f_2(\underline{\underline{X}},m,c)$ to establish Theorem 5 for arbitrary c.

Thus Theorem 5 follows from the case c=1 and so the proof of this case is our objective from now on. Section 2 contains various preliminary results and then in section 3 the proof of Theorem 5 in case c=1 is reduced to its precise statement (involving defects) when G is finite and even a p-group. A ring-theoretic description of some products of two abelian subgroups (due to Sysak [6]) is given in section 4 and then applied to a special case of Theorem 5 in section 5. The completion of the proof of Theorem 5 is given in section 6.

2. SOME PRELIMINARY RESULTS

Abelian-by-finite groups will play an important part in our arguments and we begin with the elementary

LEMMA 1. Let $G=HK=AX$ with $A \lhd G$, A abelian, $X \lhd^m H$ and $X \lhd^m K$. Then $X \lhd^{2m} G$.

Proof. Put $N=A \cap H$. Then $N \lhd AH=G$. Also $X \lhd^m NX$ ($\leq H$). Therefore if we assume that $A \cap H=1$, then it is sufficient to show that $X \lhd^m G$. In this case $H=H \cap AX=X \leq K$ and so $G=K$. The Lemma follows. □

Next a useful reduction step in our arguments will be provided by

LEMMA 2. Let $G=HK$ with $X \lhd^m H$, $X \lhd^m K$ and let $A \lhd G$. Write $G_1=AH\cap AK$, $H_1=H\cap AK$, $K_1=AH\cap K$. Then

(i) $G_1=H_1K_1=AH_1=AK_1$ and $X \lhd^m H_1$, $X \lhd^m K_1$. (Observe that $A\cap H_1=A\cap H$ and $A\cap K_1=A\cap K$.)

Now suppose in addition that A is abelian and let $N=(A\cap H)(A\cap K)$. Then

(ii) $N \lhd G_1$.

Let bars denote subgroups of G_1 modulo N. Then

(iii) $\overline{G}_1=\overline{H}_1\overline{K}_1=\overline{A}\overline{H}_1=\overline{A}\overline{K}_1$, $\overline{X} \lhd^m \overline{H}_1$, $\overline{X} \lhd^m \overline{K}_1$ and $\overline{A}$ is an abelian normal subgroup of $\overline{G}_1$. Also

$$\overline{A}\cap\overline{H}_1=\overline{A}\cap\overline{K}_1=1$$

and $\overline{H}_1$, $\overline{K}_1$ are both embeddable in G/A.

Finally suppose as a further hypothesis that $AX \lhd^{m_1} G$ and $\overline{X} \lhd^{m_2} \overline{G}_1$. Then

(iv) $X \lhd^{2m+m_1+m_2} G$.

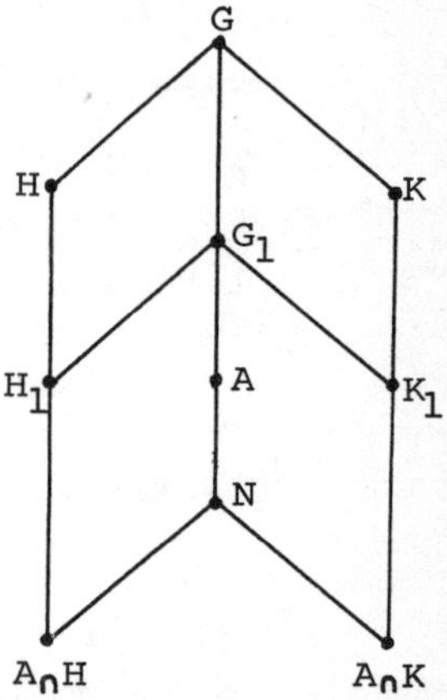

<u>Proof</u>. (i) Using Dedekind's Intersection Lemma we have

$$G_1=AH\cap AK\cap HK=HK_1\cap AK=H_1K_1.$$

Also

$$AH_1=A(H\cap AK)=AH\cap AK=G_1$$

and similarly $AK_1=G_1$. Moreover $X \lhd^m H_1$, $X \lhd^m K_1$ since $X \leqslant H_1 \leqslant H$ and $X \leqslant K_1 \leqslant K$.

(ii) Since $A\cap H=A\cap H_1 \lhd AH_1=G_1$ and similarly $A\cap K \lhd G_1$, it follows that $N \lhd G_1$.

(iii) The first statements follow trivially from (i). Also again by Dedekind's Lemma

$$A\cap NH_1=N(A\cap H_1)=N.$$

Therefore $\overline{A}\cap\overline{H}_1=1$ and similarly $\overline{A}\cap\overline{K}_1=1$. Thus $\overline{H}_1\cong\overline{G}_1/\overline{A}\cong G_1/A \leqslant G/A$ and likewise $\overline{K}_1$ embeds in G/A.

(iv) By hypothesis $NX \lhd^{m_2} G_1$ and so $NX \lhd^{m_2} AX \lhd^{m_1} G$. Therefore

$$NX \lhd^{m_1+m_2} G. \tag{1}$$

Now $NX=(A\cap H)(A\cap K)X$. Thus if $H_2=(A\cap H)X$ and $K_2=(A\cap K)X$, then $NX=H_2K_2$ and $X \lhd^m H_2$, $X \lhd^m K_2$. Therefore, by Lemma 1, $X \lhd^{2m} NX$ and (iv) follows from

(1) above. $\square$

As a first corollary we see that Theorem 1 holds for metabelian groups G.

COROLLARY 1. Let G=HK be a metabelian group with $X \lhd^m H$, $X \lhd^m K$. Then $X \lhd^{2(m+1)} G$.

<u>Proof</u>. Let $A \lhd G$ with A and G/A abelian. Then $AX \lhd G$ and so $m_1=1$ in the notation of Lemma 2. Also $\bar{G}_1=\bar{H}_1\bar{K}_1$ with $\bar{H}_1$ and $\bar{K}_1$ abelian. Therefore $\bar{X}$ lies in the centre of $\bar{G}_1$ and $m_2=1$. Thus by Lemma 2 (iv), $X \lhd^{2(m+1)} G$. $\square$

A second corollary will give us one of the hypotheses of (iv) in a later application of the Lemma.

COROLLARY 2. Suppose that G=HK with $A \lhd G$, A abelian, $|G/A| \leq n$ (finite) and $X \lhd^m H$, $X \lhd^m K$. Then $X \lhd^{\ell} G$ where $\ell = 2m+n+n^2$.

<u>Proof</u>. By Theorem 1 we know that AX is subnormal in G and so in the notation of Lemma 2 we can take $m_1=n$. Also $\bar{G}_1=\bar{H}_1\bar{K}_1$ is of finite order $\leq n^2$ and so again by Theorem 1 we have $\bar{X}$ subnormal in $\bar{G}_1$. Therefore we can take $m_2=n^2$. Then the Corollary follows from Lemma 2. $\square$

Concerning factorised groups in general we shall need

LEMMA 3 (See Amberg [1]). Let a group G be the product of subgroups H and K and suppose that $H_0 \leq H$, $K_0 \leq K$ with $|H:H_0|=r$, $|K:K_0|=s$, r and s finite. If $J=\langle H_0, K_0\rangle$, then $|G:J| \leq rs$.

<u>Proof</u>. There are elements $h_1,\ldots,h_r$ in H and $k_1,\ldots,k_s$ in K such that

$$G = HK = \bigcup_{i,j} h_i H_0 K_0 k_j = \bigcup_{i,j} h_i J k_j = \bigcup_{i,j} h_i k_j J^{k_j}. \qquad (2)$$

By a result of B. H. Neumann [4] we can omit from the right hand term of (2) those cosets for which $|G:J^{k_j}|$ is infinite. Thus for some j,

$|G:J^kJ|$ is finite and hence $|G:J|$ is finite. Denote the core of J in G by J_G. Then G/J_G is finite and therefore we may assume that G is finite (factoring by J_G). Thus $|G| \leqslant rs|J|$ by (2), whence $|G:J| \leqslant rs$. □

Sysak in [6] gives a useful description of triply factorised groups like $\overline{G}_1$ in Lemma 2 and this will enable us to reduce the proof of Theorem 5 to the finite case in section 3. Thus consider a group

$$G=HK=AH=AK$$

with $A \lhd G$, A abelian and $A \cap H = A \cap K = 1$. Let g be an element of G. Then we can write uniquely $g=ah=bk$ where $a,b \in A$, $h \in H$ and $k \in K$. Define a map $\phi : G \to G$ by

$$g^{\phi}=ak. \tag{3}$$

It is easily checked that ϕ is an automorphism of G stabilising the series $1 \lhd A \lhd G$, in particular fixing the elements of H and K modulo A. Also the fixed-point subgroup for ϕ is $A(H \cap K)$. Thus $H^{\phi}=K$ and the subgroup of elements of H (or K) which are fixed by ϕ is $H \cap K$. Now form the semidirect product

$$\Gamma = G \rtimes \langle \phi \rangle .$$

So $\Gamma = (AH) \rtimes \langle \phi \rangle = (A \times \langle \phi \rangle) \rtimes H$.

Let $a \in A$. Then $a=kh$, $h \in H$, $k \in K$, and $h^{-1}=a^{-1}k$. So $(h^{-1})^{\phi}=k$, i.e. $a=h^{-\phi}h=\phi^{-1}\phi^{h}$. Thus $\phi a = \phi^{h}$ and hence $\phi A \subseteq \phi^{H}$ (the <u>set</u> of conjugates of ϕ by elements of H). On the other hand $\phi^{H} \subseteq \phi A$ since $[\phi, H] \leqslant A$. Therefore

$$\phi A = \phi^{H}. \tag{4}$$

Conversely consider a group $\Gamma = (A \times \langle \phi \rangle) \rtimes H$ where $A^{H}=A$ and $\phi A = \phi^{H}$. Let $K=H^{\phi}$. So $K \leqslant AH = G$ say. Also if $a \in A$, then there exists $h \in H$ such that $a=\phi^{-1}h^{-1}\phi h \in KH$. Therefore $A \leqslant KH$ and hence $G=AH \leqslant KH \leqslant G$. Thus

$$G=HK. \tag{5}$$

This completes the preliminaries necessary for the proof of Theorem 5.

3. REDUCTION OF THEOREM 5 TO THE FINITE CASE

We have already seen that it is necessary to consider only groups G which have finite order $\leqslant n$ modulo a normal metabelian subgroup. Thus let $A \lhd G$, A abelian, $A \leqslant B \lhd G$, B/A abelian and $|G/B| \leqslant n$. We have

$G=HK$ with $X \lhd^m H$, $X \lhd^m K$ and G is periodic. In order to prove that X is subnormal in G with defect bounded by a function of m and n, we show that it suffices to establish the result when G is finite.

By Lemma 2 and its Corollary 2 we may assume that

$$\underline{G=HK=AH=AK \text{ and } A\cap H=A\cap K=1.} \tag{6}$$

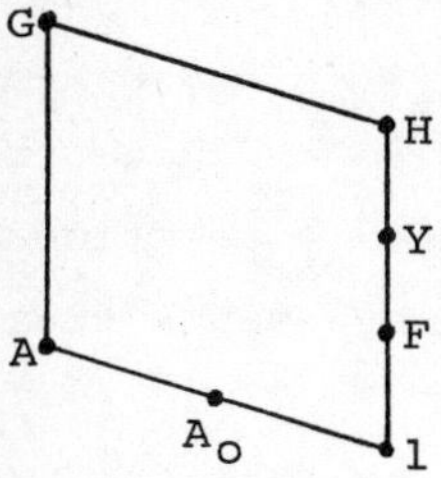

Let F be a normal subgroup of H and let $A_0=[\phi,F]$, where ϕ is given by (3). Following Sysak [6] define

$$Y=\{y\in H \mid \phi^y\in \phi A_0\}.$$

It is easy to check that the product of any two elements of Y also lies in Y. Therefore since G is periodic, Y is a subgroup of H containing F. (This is the point at which we require the hypothesis that G is periodic, otherwise it would not be clear that Y is a subgroup. If Y were a subgroup in general, then the following argument can be modified to show that Theorem 5 can be reduced to the finite case even when G is not periodic.)

We have $\phi^Y \subseteq \phi A_0$. Conversely let $a\in A_0$. By (4) there is an element $h\in H$ such that $a=\phi^{-1}\phi^h$. Therefore $\phi^h=\phi a\in\phi A_0$ and so $h\in Y$ by definition of Y. Thus $\phi a\in\phi^Y$ and hence $\phi A_0 \subseteq \phi^Y$. Therefore $\phi A_0=\phi^Y$. Now Y normalises A_0 and if $\Gamma=\langle A_0,\phi,Y\rangle$, then

$$\Gamma=(A_0\times\langle\phi\rangle)\rtimes Y.$$

Therefore if $G_0=A_0Y$, (5) shows that $G_0=YZ$ where $Z=Y^\phi\leqslant K$. Also $X\leqslant H\cap K$ implies $[\phi,X]=1$ and so $X\leqslant Y\cap Z$, i.e. $X\lhd^m Y$, $X\lhd^m Z$.

Now suppose that F is a finite normal subgroup of H. It is clear that G is a locally finite group and therefore $\langle F,F^\phi\rangle$ is finite. It follows that $A_0=[\phi,F]$ is finite and hence $|G_0:Y|$ is finite. Suppose that Theorem 5 has been proved for finite groups. Then there is an integer $f_3=f_3(m,n)$ such that if Y_0 is the core of Y in G_0, we have $Y_0X\lhd^{f_3}G_0$. But $Y_0X\leqslant H$ and so $X\lhd^m Y_0X$. Therefore $X\lhd^{m+f_3}G_0$ and hence

$$[A_0,{}_{m+f_3}X]=1. \tag{7}$$

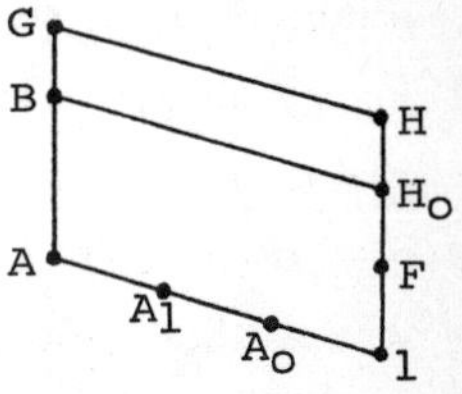

Let $H_0=B\cap H$. Then $H_0\lhd H$, H_0 is abelian and $|H/H_0|\leqslant n$. Now H_0 is generated by finite normal subgroups F of H. Thus $A_1=[\phi,H_0]$ is generated by the subgroups $[\phi,F]$ $(=A_0)$. Hence by (7) $[A_1,{}_{m+f_3}X]=1$. Since X normalises each A_0, X

also normalises A_1. Clearly H_o normalises A_1 and therefore $A_1X \lhd^m A_1H_oX$. Then $X \lhd^{2m+f_3} A_1H_oX$. However, $H_o^{\phi}=K_o$ (say) is contained in A_1H_oX and so

$$X \lhd^{2m+f_3} \langle H_o, K_o, X\rangle. \qquad (8)$$

We have $|K:K_o| \leqslant n$ and hence, by Lemma3, $J=\langle H_o, K_o\rangle$ has finite index $\leqslant n^2$ in G. Therefore G/J_G has finite order bounded by a function of n, and since XJ_G is subnormal in G (by Theorem 1), XJ_G has defect in G bounded by a function of n. But $X \lhd^{2m+f_3} XJ_G$ by (8) and hence it follows that X is subnormal in G with defect bounded by a function of m and n. Thus to complete the proof of Theorem 5 we may assume that G is finite (and of course that B is metabelian).

Finally in this section we show that we may assume that

$$G \text{ is a p-group} \qquad (9)$$

for some prime p. For certainly we may assume as before that the relations (6) hold and therefore that A is a p-group. Let q be any prime different from p and let X_q be any Sylow q-subgroup of X. Put $F=AX_q$. We know by Theorem 1 that X is subnormal in G and hence $X \cap F = X_q$ is subnormal in F. Therefore $[A, X_q]=1$ and so if W is the p-residual of X, then $[A,W]=1$. Factoring G by W^G ($\leqslant H \cap K$), we may assume that X is a p-group.

By a result of Wielandt [7] there is a Sylow p-subgroup G_p of G such that

$$G_p = H_pK_p$$

where H_p, K_p are Sylow p-subgroups of H, K respectively. Note that $X \lhd^m H_p$, $X \lhd^m K_p$ and $A \leqslant G_p$. Since it suffices to show that X has subnormal defect in AX bounded by a function of m and n, we can take G_p for G, i.e. (9) follows.

4. A RING-THEORETIC DESCRIPTION OF SOME PRODUCTS OF ABELIAN GROUPS

When we return to the proof of Theorem 5 we have to consider certain products of abelian groups. Our arguments will be facilitated by making use of a representation of these products as groups of 2×2 matrices over certain rings. (See Sysak [6]. The author also wishes to acknowledge the opportunity of seeing an unpublished manuscript by

R. B. Howlett containing the same results as [6], though prepared independently of that paper.)

Consider a group $G=UV=AU=AV$ where A,U,V are abelian subgroups of G, $A \lhd G$ and $U\cap V=A\cap U=A\cap V=1$. For each g in G we can write uniquely $g=u^{-1}v$ with u in U, v in V. In particular each $a\in A$ is uniquely expressible as $a=u_a^{-1}v_a$, u_a in U, v_a in V. Let $\alpha:A\to U$ be the map defined by

$$a^{\alpha}=u_a.$$

Given $u\in U$, there are unique elements $a\in A$, $v\in V$ such that $u^{-1}=av^{-1}$. Then $a=u^{-1}v$ and hence α is a bijection.

Let R be a set such that there is a bijection

$$\theta:A\to R.$$

We give R an additive group structure isomorphic to the group A by defining $\theta(a)+\theta(b)=\theta(ab)$, all a,b in A. Now define a binary relation $*$ on A as follows: $a*b=[a,b^{\alpha}]$ for all a,b in A. It is routine to check that $*$ is commutative and associative. Next define a multiplication on R by $\theta(a).\theta(b)=\theta(a*b)$ for all a,b in A. Again it is routine to check that this multiplication is distributive over the addition of R and R becomes an associative and commutative ring.

Embed R in a ring S with 1 in the usual way. Then there is a bijection

$$\psi:U\to 1+R \quad (=\{1+r \mid r\in R\})$$

given by $\psi(a^{\alpha})=1+\theta(a)$ $(a\in A)$ and it is easy to check that ψ preserves multiplication, i.e.

<u>$1+R$ is a multiplicative group isomorphic to U.</u>

Moreover if G_R is the multiplicative group of matrices $\begin{pmatrix}1+r & 0\\ s & 1\end{pmatrix}$ (all r,s in R), then $G\cong G_R$ via the map

$$\gamma: a^{\alpha}b \longrightarrow \begin{pmatrix}1+\theta(a) & 0\\ \theta(b) & 1\end{pmatrix}$$

(a,b in A). If

$$U_R=\left\{\begin{pmatrix}1+r & 0\\ 0 & 1\end{pmatrix} \middle| \; r\in R\right\},\quad V_R=\left\{\begin{pmatrix}1+r & 0\\ r & 1\end{pmatrix} \middle| \; r\in R\right\} \text{ and } A_R=\left\{\begin{pmatrix}1 & 0\\ r & 1\end{pmatrix} \middle| \; r\in R\right\},$$

then $U^{\gamma}=U_R$, $V^{\gamma}=V_R$ and $A^{\gamma}=A_R$.

We choose to view G_R as a group on $R\times R$ with multiplication defined by

$$(r_1,s_1)(r_2,s_2)=(r_1+r_2+r_1r_2,\, s_1+s_2+s_1r_2),$$

all r_i,s_i in R. Then

$U^\gamma = \{(r,0) \mid r\in R\}$, $V^\gamma = \{(r,r) \mid r\in R\}$, $A^\gamma = \{(0,r) \mid r\in R\}$.

Shortly we shall have to consider the subgroup X of Theorem 5 acting on certain products of abelian subgroups. In preparation for this, and with the former notation, consider a subgroup X of AutG with $U^X=U$, $V^X=V$, $A^X=A$. Then

X induces (in a natural way) a group of automorphisms of R. (10)

For, via γ, X acts as a group of automorphisms of G_R. In particular for $x\in X$ we have a map $\lambda:R\to R$ defined by $(0,r)^x=(0,r^\lambda)$, $r\in R$. Then clearly λ is an automorphism of the additive group of R.

Given (r,r) in V_R, there exists $(s,0)$ in U_R such that $(s,0)(r,r)\in A_R$, since $V_R \leq U_R A_R$ $(=G_R)$. Then

$$(s,0)(r,r)=(0,r). \qquad (11)$$

With the same x as above, write $(r,r)^x=(r^\mu,r^\mu)$. Then $\mu:R\to R$ is a bijection. Applying x to (11), there is an element s_1 in R such that $(s_1,0)(r^\mu,r^\mu)=(0,r^\lambda)=(s_1+r^\mu+s_1r^\mu,r^\mu)$. Therefore $\mu=\lambda$ and $(r,r)^x=(r^\lambda,r^\lambda)$.

Now for any r, s in R

$$((r,r)(s,s))^x=(r^\lambda,r^\lambda)(s^\lambda,s^\lambda)=(r^\lambda+s^\lambda+r^\lambda s^\lambda,r^\lambda+s^\lambda+r^\lambda s^\lambda)$$
$$=(r+s+rs,r+s+rs)^x=(r^\lambda+s^\lambda+(rs)^\lambda,r^\lambda+s^\lambda+(rs)^\lambda).$$

Thus $(rs)^\lambda=r^\lambda s^\lambda$ and hence λ is an automorphism of R. Then (10) follows.

We also have

$$(r,s)^x=(r^\lambda,s^\lambda) \qquad (12)$$

for all r, s in R. For, write $(r,0)^x=(r^\nu,0)$, all $r\in R$. So $\nu:R\to R$ is a bijection and $((r,0)(0,r))^x=(r,r)^x=(r^\lambda,r^\lambda)=(r^\nu,0)(0,r^\lambda)=(r^\nu,r^\lambda)$. Hence $\nu=\lambda$ and $(r,s)^x=((r,0)(0,s))^x=(r^\lambda,0)(0,s^\lambda)=(r^\lambda,s^\lambda)$. Thus (12) follows.

Conversely it is very easy to show that if λ is an automorphism of R, then the map $x:G_R\to G_R$ defined by $(r,s)^x=(r^\lambda,s^\lambda)$ is an automorphism of G_R and U_R, V_R, A_R are all x-invariant.

Observe that the action of x on U (and on V) is isomorphic to the action of λ on the multiplicative group 1+R (with $(1+r)^\lambda=1+r^\lambda$); and the action of x on A is isomorphic to the action of λ on the additive group of R.

5. THE RING-THEORETIC INGREDIENT OF THE PROOF OF THEOREM 5

Consider a group G which is the product of three subgroups

$$\left.\begin{array}{l} G=UVX \text{ with } U,V \text{ abelian, } UV=VU, \\ U^X=U,\ V^X=V \text{ and } X \triangleleft^m UX,\ X \triangleleft^m VX. \end{array}\right\} \qquad (13)$$

Our interest lies in the problem of bounding the defect of X whenever it is subnormal in G. Accordingly our first objective is to relate $J=UV$ to a group with the structure described in section 4.

By Itô's Theorem, J is metabelian. Let $A=J'$. Then $A^X=A$. Put $J_1=AU\cap AV$, $U_1=U\cap AV$, $V_1=AU\cap V$. Then X normalises J_1, U_1 and V_1 and $J_1=U_1V_1=AU_1=AV_1$ by Lemma 2 (i). Also $X \triangleleft^m U_1X$, $X \triangleleft^m V_1X$ and $A\cap U=A\cap U_1$, $A\cap V=A\cap V_1$. Let $N=(A\cap U)(A\cap V)$. By Lemma 2 (ii), $N \triangleleft J_1$. If bars denote subgroups of J_1 modulo N, then by Lemma 2 (iii)

$$\bar{J}_1=\bar{U}_1\bar{V}_1=\bar{A}\bar{U}_1=\bar{A}\bar{V}_1,$$

$\bar{U}_1$ and $\bar{V}_1$ are abelian, $\bar{A}$ is an abelian normal subgroup of $\bar{J}_1$ and

$$\bar{A}\cap\bar{U}_1=\bar{A}\cap\bar{V}_1=1.$$

Moreover X normalises N and with $\bar{X}=XN/N$ we have $\bar{X} \triangleleft^m \bar{U}_1\bar{X}$, $\bar{X} \triangleleft^m \bar{V}_1\bar{X}$.

Now J/A is abelian and so Lemma 1 shows that $AX \triangleleft^{2m} G$. Suppose that

$$\bar{X} \triangleleft^{\ell} \bar{J}_1\bar{X}. \qquad (14)$$

Then as in Lemma 2 (iv) we see that

$$\left.\begin{array}{c} \underline{\text{X is subnormal in G with defect bounded}} \\ \underline{\text{by a function of } \ell \text{ and m.}} \end{array}\right\} \qquad (15)$$

We now focus our attention on a case where (14) holds.

LEMMA 4. Consider a finite group $G=UVX$ where U,V are abelian subgroups and X is a subgroup of order $\leq n$ normalising U and V. Suppose that there is a normal abelian subgroup A of G such that

$$UV=VU=AU=AV \text{ and } A\cap U=A\cap V=1.$$

Let $X \triangleleft^m UX$, $X \triangleleft^m VX$. Then there is an integer $f_4=f_4(m,n)$ such that $X \triangleleft^{f_4} G$.

Thus starting with the hypotheses (13) where G is finite and $|X| \leq n$, Lemma 4 shows that (14) holds with ℓ bounded by a function of m and n. Then (15) follows.

<u>Proof of Lemma 4</u>. Observe that $U\cap V$ lies in the centre of $J=UV$

and X normalises $U \cap V$. Also $X \lhd^m (U \cap V)X$ ($\leqslant UX$). Therefore we may factor G by $U \cap V$ and assume that $U \cap V=1$.

Now $[U,{}_mX] \leqslant U \cap X \leqslant C_X(U)=C$ (say) since U is abelian. Moreover

$$U \overset{X}{\cong} J/A \overset{X}{\cong} V$$

and therefore $C=C_X(V)$. Thus $C=C_X(J) \lhd G$ and hence we may assume that $C=1$, i.e.

$$\left.\begin{array}{l}\underline{\text{X stabilises a series of U of length m}} \\ \text{and} \quad \underline{\text{X is nilpotent of class} \leqslant \text{m-1}}\end{array}\right\} \qquad (16)$$

by a well-known result of Philip Hall [2]. We may lose the hypotheses $A \cap U=A \cap V=1$ by taking $C=1$, but they can be retrieved by arguing as we did above from (13) and without losing (16). Similarly $U \cap V=1$ can also be retrieved.

Clearly it suffices to show that there is an integer $f_5=f_5(m,n)$ such that

$$\underline{\text{X stabilises a series of A of length } f_5}. \qquad (17)$$

For this purpose there is no loss in identifying X with the group of automorphisms of J which it induces by conjugation.

We have seen in section 4 that there is an associative and commutative ring R such that

$$J \overset{X}{\cong} J_R$$

and X acts as a group of automorphisms of R. From (16) we see that

$$\left.\begin{array}{c}\underline{\text{X stabilises a series of the multiplicative}} \\ \underline{\text{group } 1+R \ (\overset{X}{\cong}U) \text{ of length m.}}\end{array}\right\} \qquad (18)$$

To prove (17) we must show that

$$\left.\begin{array}{c}\underline{\text{X stabilises a series of the additive}} \\ \underline{\text{group R } (\overset{X}{\cong}A) \text{ of length } f_5.}\end{array}\right\} \qquad (19)$$

The remainder of the argument is ring-theoretic and the greater part of it is not restricted to the situation where G and therefore R are finite, nor where X is finite. Therefore it is convenient to interrupt the proof of Lemma 4 at this point in order to establish a general result about the ring R and a group X of automorphisms of R satisfying (18).

Let S be an associative ring with 1 and let R be an ideal of S with the property that each element of the form $1+r$ ($r \in R$) is a unit. Then R is called a _radical_ ring and the set $1+R$ is a multiplicative group. For if $(1+r)s=1$ for some $s\in S$, then $s=1-rs \in 1+R$. We have

LEMMA 5. Let X be a group of automorphisms of the commutative radical ring R and suppose that X stabilises a series of the multiplicative group 1+R of length m. Then there is a positive integer $h=h(m)$ such that X stabilises a series of the additive group hR of length $\leqslant m$. In fact with $t=2^{m-1}$,

$$h=2^{t-1}\prod_{i=1}^{t-1}(2^i-1).$$

<u>Proof</u>. By hypothesis, for all r in R and $x_1,\ldots,x_m \in X$ we have

$$(1+r)^{(x_1-1)\ldots(x_m-1)}=1.$$

Now

$$\prod_{i=1}^{m}(x_i-1)=x_1\ldots x_m-\Sigma x_{i_1}\ldots x_{i_{m-1}}+\Sigma x_{i_1}\ldots x_{i_{m-2}}-\ldots+(-1)^m$$

with the obvious summation ranges. Therefore if

$$\xi_1=x_1\ldots x_m+\Sigma x_{i_1}\ldots x_{i_{m-2}}+\ldots$$

and

$$\xi_2=\Sigma x_{i_1}\ldots x_{i_{m-1}}+\Sigma x_{i_1}\ldots x_{i_{m-3}}+\ldots,$$

it follows that $(1+r)^{\xi_1}=(1+r)^{\xi_2}$. Hence

$$(1+rx_1\ldots x_m).\Pi(1+rx_{i_1}\ldots x_{i_{m-2}})\ldots$$
$$=\Pi(1+rx_{i_1}\ldots x_{i_{m-1}}).\Pi(1+rx_{i_1}\ldots x_{i_{m-3}})\ldots$$

and we have $t=2^{m-1}$ factors on each side. (Note that when we pass from considering the multiplicative group 1+R as an X-module, we lower the operators x_i from the superscript position to the same line as the elements of R.) Expanding each side of the above equation and collecting the terms of degree 1 in the elements of R on the left side gives

$$rx_1\ldots x_m-r\Sigma x_{i_1}\ldots x_{i_{m-1}}+\ldots+(-1)^m r$$
$$=r(x_1-1)\ldots(x_m-1)=\sum_{i=2}^{t}\sum_{j=1}^{2\binom{t}{i}}\prod_{k=1}^{i} ry_{i,j,k} \qquad (20)$$

where each $y_{i,j,k}$ is a monomial in the x_i's independent of r.

Now replace r by 2r in (20). Then

$$2r(x_1-1)\dots(x_m-1)=\sum_{i=2}^{t}2^i\sum_{j=1}^{2\binom{t}{i}}\prod_{k=1}^{i}ry_{i,j,k}. \tag{21}$$

Multiplying (20) by 2^2 and subtracting (21) gives

$$2(2-1)r(x_1-1)\dots(x_m-1)=\sum_{i=3}^{t}(2^2-2^i)\sum_{j=1}^{2\binom{t}{i}}\prod_{k=1}^{i}ry_{i,j,k} \tag{22}$$

and the right hand side has degree $\geqslant 3$ as a polynomial in the elements $ry_{i,j,k}$. Again replacing r by 2r in (22) and subtracting the result from (22) multiplied by 2^3 gives

$$2^2(2^2-1)(2-1)r(x_1-1)\dots(x_m-1)=\sum_{i=4}^{t}(2^3-2^i)(2^2-2^i)\sum_{j=1}^{2\binom{t}{i}}\prod_{k=1}^{i}ry_{i,j,k}$$

and the right hand side has degree $\geqslant 4$. Continuing in this way we find that

$$2^{t-1}(2^{t-1}-1)(2^{t-2}-1)\dots(2-1)r(x_1-1)\dots(x_m-1)=0$$

and this holds for all $r\in R$ and $x_1,\dots,x_m\in X$. Thus $hR(x_1-1)\dots(x_m-1)=0$, i.e. writing $\underline{\underline{x}}$ for the augmentation ideal of the group ring $\mathbf{Z}X$, $hR\underline{\underline{x}}^m=0$. Therefore X stabilises a series of hR of length $\leqslant m$. □

<u>Proof of Lemma 4 continued</u>. We have to deduce (19) from (18). Since G is finite and X is nilpotent, we know from Theorem 1 that X stabilises a series of A. Then since the additive group of R is X-isomorphic to A, it follows that X stabilises a series of R.

By Lemma 5 we may assume that $hR=0$. Recall that $|X|\leqslant n$. Thus if $r\in R$, then the additive group $\langle rX\rangle$ has order $\leqslant h^n$. Therefore X stabilises a series of $\langle rX\rangle$ and hence of R of length $\leqslant h^n$ and (19) follows. This completes the proof of Lemma 4. □

We have shown that a finite group G satisfying (13) with $|X|\leqslant n$ contains X as a subnormal subgroup with defect bounded by a function of m and n. For later reference we record this as

LEMMA 6. Let U,V,X be subgroups of a finite group G with $G=UVX$, U and V abelian, $UV=VU$, $U^X=U$, $V^X=V$ and $X\lhd^m UX$, $X\lhd^m VX$. If $|X|\leqslant n$, then X is subnormal in G with defect bounded by a function of m and n.

6. COMPLETION OF THE PROOF OF THEOREM 5

In section 3 it was shown that it suffices to prove Theorem 5 in the case when $G=HK$ is a finite p-group, $B\lhd G$, B is metabelian and $|G/B|\leqslant n$. We have $X\lhd^m H$, $X\lhd^m K$ and must show that the subnormal defect of X in G is bounded by a function of m and n.

Let $A\lhd G$ with $A\leqslant B$ where A and B/A are abelian. By means of Lemma 2 and its Corollary 2, we can assume (as at the beginning of section 3; see (6)) that

$$G=HK=AH=AK,\ A\cap H=A\cap K=1.$$

Put $H_o=B\cap H$, $K_o=B\cap K$, $J=\langle H_o,K_o\rangle$. Since $|H:H_o|=|K:K_o|\leqslant n$, we see from Lemma 3 that $|G:J|\leqslant n^2$. Let $D=\bigcap_{k\in K} J^k$. Then $|G:D|\leqslant n^{2n}$. Let $K_o\leqslant K_1\leqslant \dots \leqslant K_d=D$ be the normal closure series of K_o in D. Since K normalises D and K_o, K also normalises each subgroup K_i.

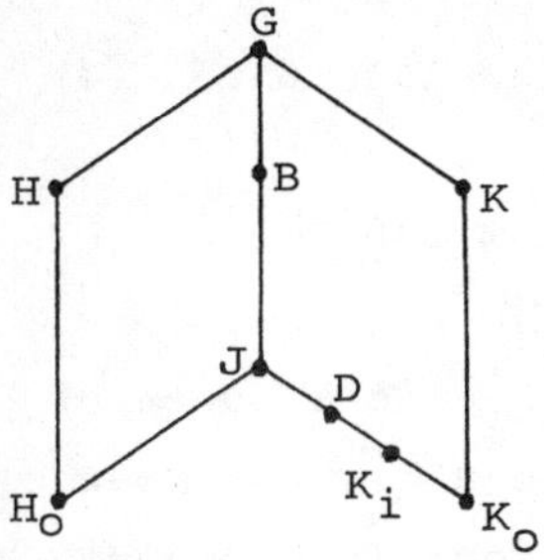

Choose i maximal such that $K_i\subseteq H_oK_o$. So $0\leqslant i\leqslant d$ and $K_i=(H_o\cap K_i)K_o$. The factors on the right are both abelian and normalised by X. Also $X_o=X\cap H_o\leqslant K\cap B=K_o$ and and therefore X_o lies in the centre of J. Moreover $X/X_o\cong XH_o/H_o\leqslant H/H_o$ and hence $|X/X_o|\leqslant n$. Now X acts (by conjugation) on K_i and is subnormal of defect $\leqslant m$ in its join with each of $H_o\cap K_i$ and K_o. Therefore by Lemma 6 X/X_o has defect in its join with K_i/X_o bounded by a function of m and n. Thus

<u>the defect of X in XK_i is bounded by a function of m and n.</u> (23)

If $i=d$, then $K_i=D$ and since $|G:D|\leqslant n^{2n}$, the subnormal defect of X in G is bounded by a function of m and n. Therefore suppose that $i<d$. We claim that

<u>the defect of X in XK_{i+1} is bounded by a function of m and n.</u> (24)

To see this let $N=N_G(K_i)$. Then $K\leqslant N$ and so $N=(N\cap H)K$. Now $XK_i\lhd^m (N\cap H)K_i$ since $X\lhd^m H$. Therefore by (23)

<u>the defect of X in $(N\cap H)K_i$ is bounded by a function of m and n.</u> (25)

But $|N:(N\cap H)K_i|=|(N\cap H)K:(N\cap H)K_i|=|K:(N\cap H)K_i\cap K|\leqslant |K:K_o|\leqslant n$. Thus from (25) it follows that the defect of X in N is bounded by a function of m and n. Since $XK_{i+1}\leqslant N$, (24) holds. Note that $H_oK_o\subsetneqq H_oK_{i+1}$ since

$K_{i+1} \nsubseteq H_oK_o$.

Now choose j maximal such that $K_j \subseteq H_oK_{i+1}$. Then $i+1 \leqslant j \leqslant d$ and $K_j=(H_o \cap K_j)K_{i+1}$. We claim that

<u>the defect of X in XK_j is bounded by a function of m and n</u>. (26)

For, $A \cap K_{i+1}=A_{i+1}$ (say) is normal in AK=G. Since X has bounded defect in $A_{i+1}X$ (by (24)), it suffices to show that X has bounded defect in XK_j modulo A_{i+1}. Let bars denote subgroups modulo A_{i+1}. Then

$$\overline{K}_{i+1} \cong AK_{i+1}/A \leqslant B/A$$

and so $\overline{K}_{i+1}$ is abelian. Also X acts on

$$\overline{K}_j = (\overline{H_o \cap K_j})\overline{K}_{i+1}$$

and fixes both of the abelian factors on the right hand side. Therefore by Lemma 6, $\overline{X}/\overline{X}_o$ has bounded defect in its join with $\overline{K}_j/\overline{X}_o$ and thus $\overline{X}$ has bounded defect in its join with $\overline{K}_j$. Then (26) follows.

If j=d we finish as before. Otherwise, by analogy with (24),

X has bounded defect in XK_{j+1}.

Also $H_oK_{i+1} \subsetneqq H_oK_{j+1}$, since $K_{j+1} \nsubseteq H_oK_{i+1}$. Now from G=HK we have

$$|G||H \cap K| = |H||K| \leqslant n^2 |H_o||K_o| = n^2 |H_oK_o||H_o \cap K_o| .$$

Therefore $|G|/|H_oK_o| \leqslant n^2$. Since, for all i, $|H_oK_i|$ is a power of p, it follows that the above process must terminate after a number of steps bounded by a function of n. Then X has defect in G bounded by a function of m and n and this completes the proof of Theorem 5. □

• It would be interesting to know if Theorem 5 requires the periodicity of G. Observe that Theorems 3 and 4 do not. Certainly Theorem 5 is true for any group G which is the extension of a nilpotent group A by a periodic abelian-by-finite group. For, using the arguments which we have already employed we may assume that A is abelian and then by Lemma 2 that H and K embed in G/A, i.e. H and K are periodic. But then Sysak shows in [6] that G=HK is periodic and therefore the argument which we have presented applies.

REFERENCES

1 B. Amberg, Artinian and noetherian factorized groups, Rend. Sem. Mat. Univ. Padova 55 (1976), 105-122.

2 P. Hall, Some sufficient conditions for a group to be nilpotent, Illinois J. Math. 2 (1958), 787-801.

3 R. Maier, Um problema da teoria dos subgrupos subnormais, Bol. Soc. Brasil Mat. 8 (1977), 127-130.

4 B.H. Neumann, Groups covered by permutable subsets, J. London Math. Soc. 29 (1954), 236-248.

5 S.E. Stonehewer, Subnormal subgroups of products of groups (to appear).

6 Ya.P. Sysak, Products of infinite groups, preprint no. 82.53, Akad. Nauk Ukr. SSR, Inst. Mat. Kiev (1982).

7 H. Wielandt, Über das Produkt paarweise vertauschbarer nilpotenter Gruppen, Math. Z. 55 (1951), 1-7.

8 H. Wielandt, Über das Erzeugnis paarweise kosubnormaler Untergruppen, Arch. Math. 35 (1980), 1-7.

9 H. Wielandt, Subnormalität in faktorisierten endlichen Gruppen, J. Algebra 69 (1981), 305-311.

AN EMBEDDING CONDITION FOR SUBGROUPS OF INFINITE GROUPS

John S. Wilson
Department of Pure Mathematics and Mathematical Statistics
University of Cambridge
Cambridge CB2 1SB, England

Let H be a subgroup of a group G. We shall say that H is *elliptically embedded* in G, and write H ee G, if for each subgroup K of G there is an integer n depending on K such that

$$\langle H, K\rangle = HK\ldots HK,$$

where the product has 2n factors. Thus it is clear that H ee G if either G is finite or H is a quasinormal subgroup of G. A rather less trivial result, proved by P. Hall in lectures in the 1960's, is that if G is a finitely generated nilpotent group then every subgroup of G is elliptically embedded in G. A very similar argument shows that the same conclusion holds if G is finitely generated and finite by nilpotent. On the other hand, free nilpotent groups of class two and infinite rank have subgroups which are not elliptically embedded, and indeed elliptic embedding is a rather strong property. The following results have been proved in joint work with A. H. Rhemtulla.

THEOREM 1. *Suppose* $G = \langle H_1, \ldots, H_n\rangle$ *is soluble and* H_r ee G *for each* r. *If each subgroup* H_r *is finitely generated and finite by nilpotent, then so is* G.

As an immediate consequence of Theorem 1, we have

COROLLARY 1. *Let* G *be a torsion-free locally soluble group. If* H ee G *and* H *is finitely generated and nilpotent, then the normal closure* H^G *of* H *in* G *is locally nilpotent.*

This follows since each subgroup generated by finitely many conjugates of H satisfies the hypotheses of Theorem 1. However it has long been known that the normal closure of a nilpotent subnormal subgroup is locally nilpotent (see Robinson [4] p. 61), and we suspect that the subgroup H in Corollary 1 is actually subnormal. We can establish this only under much more restrictive hypotheses.

THEOREM 2. *Let* G *be a torsion-free soluble group and* H *a cyclic subgroup. If* H ee G, *then* H *is subnormal in* G.

In contrast to this result, we shall give below an example of a torsion-free metabelian group G and an elliptically embedded free abelian subgroup H of infinite rank, such that H is not subnormal in G.

Our third theorem gives a characterization of elliptically embedded subgroups of polycyclic by finite groups:

THEOREM 3. *Let* G *be polycyclic by finite and let* H *be a subgroup of* G. *Then* H ee G *if and only if* H *is subnormal in some subgroup of finite index in* G.

The close connection between subnormality and elliptic embedding which emerges in Theorem 2 and Theorem 3 is perhaps somewhat surprising. Another feature of these results which is not immediately apparent is that they depend crucially on a deep and rather recent result in number theory. The proofs make use of the following theorem, which is a consequence of results of Evertse [1] and Van der Poorten [5].

THEOREM 4. *Let* k *be an algebraic number field and let* U *be a finitely generated subgroup of the multiplicative group* $k^* = k \setminus \{0\}$ *with* $-1 \in U$. *For each integer* n *there is an element of the additive subgroup spanned by* U *which cannot be expressed as a sum of at most* n *elements of* U.

If the subgroup U in Theorem 4 is cyclic, then the result becomes much easier to prove, though still not entirely elementary. An *ab initio* treatment of this case was given in [2], and proofs of Theorem 2 and the case of Theorem 1 in which each subgroup H_r is cyclic were given. The paper [2] also contains a number of sufficient conditions for a subgroup to be elliptically embedded. The proof of Theorem 3 appears in [3].

We now explain how Theorem 4 comes to be relevant to the group-theoretic results. Let k be an algebraic number field and U a subgroup of k*, and let I be the additive subgroup generated by U. Write G for the group of matrices

$$\begin{pmatrix} u & 0 \\ z & 1 \end{pmatrix}$$

with $u \in U$ and $z \in I$, and H, A respectively for the subgroups consisting of the diagonal matrices in G and the lower unitriangular matrices in G. Thus G is a split extension of A by H. It is well known and easy to check that

(1) $H = N_G(H)$, *and* $|G : M|$ *is finite for every subgroup* M *of* G *properly containing* H.

Let K be the subgroup H^e, where e is the matrix

$$\begin{pmatrix} 1 & 0 \\ 1 & 1 \end{pmatrix} .$$

Routine but rather tedious calculations show that

(2) H ee G *if and only if* $\langle H, K\rangle = (HK)^m$ *for some integer* m,

and that

(3) $(HK)^m \cap A$ *consists of the lower unitriangular matrices with bottom left entries the elements of* I *of the form*

$$1 + \Sigma_1^{m-1} u_{2j} - \Sigma_1^m u_{2j-1} ,$$

with $u_j \in U$ *for each* j.

Now suppose that U is finitely generated and $-1 \in U$. Since $K \cong H \cong U$, the group $\langle H, K\rangle$ is generated by two finitely generated abelian subgroups which are conjugate in G, and yet is not finite by nilpotent. So by Theorem 1 we cannot have H ee G. From (2) and (3) we conclude that for each integer m there is an element of I which cannot be expressed in the form

$$1 + \Sigma_1^{2m+1} u_j ,$$

hence cannot be expressed in the form

$$\Sigma_1^{2m} u_j = 1 + (-1 + \Sigma_1^{2m} u_j) .$$

Therefore Theorem 1 implies Theorem 4.

On the other hand, if we take k to be the field of rationals and U to be the multiplicative group of positive rationals, then we have

$$(HK) \cap A \neq A = (HK)^2 \cap A ,$$

so that

$$(HK)^2 = H(HK)^2 = HA = G ,$$

and H ee G by (2). In this case G is a torsion-free metabelian group, and H is a free abelian subgroup of infinite rank which is elliptically embedded but which by (1) is not subnormal in any subgroup of finite index in G.

REFERENCES

[1] J.H. Evertse, On sums of S-units and linear recurrences, Compositio Math. 53 (1984), 225-244.

[2] A.H. Rhemtulla and J.S. Wilson, On elliptically embedded subgroups of soluble groups, Canadian J. Math.

[3] A.H. Rhemtulla and J.S. Wilson, Elliptically embedded subgroups of polycyclic groups, to appear.

[4] D.J.S. Robinson, Finiteness conditions and generalized soluble groups, Springer, Berlin-Heidelberg-New York 1972.

[5] A.J. Van der Poorten, Additive relations in number fields, Seminaire de Theorie des Nombres de Paris (1982-83), 259-266.

LECTURE NOTES IN MATHEMATICS
Edited by A. Dold and B. Eckmann

Some general remarks on the publication of proceedings of congresses and symposia

Lecture Notes aim to report new developments - quickly, informally and at a high level. The following describes criteria and procedures which apply to proceedings volumes.

1. One (or more) expert participant(s) of the meeting should act as the responsible editor(s) of the proceedings. They select the papers which are suitable (cf. points 2, 3) for inclusion in the proceedings, and have them individually refereed (as for a journal). It should not be assumed that the published proceedings must reflect conference events faithfully and in their entirety. Contributions to the meeting which are not included in the proceedings can be listed by title. The series editors will normally not interfere with the editing of a particular proceedings volume - except in fairly obvious cases, or on technical matters, such as described in points 2, 3. The names of the responsible editors appear on the title page of the volume.

2. The proceedings should be reasonably homogeneous (concerned with a limited area). For instance, the proceedings of a congress on "Analysis" or "Mathematics in Wonderland" would normally not be sufficiently homogeneous.

   One or two longer survey articles on recent developments in the field are often very useful additions to such proceedings - even if they do not correspond to actual lectures at the congress. An extensive introduction on the subject of the congress would be desirable.

3. The contributions should be of a high mathematical standard and of current interest. Research articles should present new material and not duplicate other papers already published or due to be published. They should contain sufficient information and motivation and they should present proofs, or at least outlines of such, in sufficient detail to enable an expert to complete them. Thus resumes and mere announcements of papers appearing elsewhere cannot be included, although more detailed versions of a contribution may well be published in other places later.

   Surveys, if included, should cover a sufficiently broad topic, and should in general not simply review the author's own recent research. In the case of surveys, exceptionally, proofs of results may not be necessary.

   The editors of a volume are strongly advised to inform contributors about these points at an early stage.

.../...

4. Proceedings should appear soon after the meeeting. The publisher should, therefore, receive the complete manuscript within nine months of the date of the meeting at the latest.

5. Plans or proposals for proceedings volumes should be sent to one of the editors of the series or to Springer-Verlag Heidelberg. They should give sufficient information on the conference or symposium, and on the proposed proceedings. In particular, they should contain a list of the expected contributions with their prospective length. Abstracts or early versions (drafts) of some of the contributions are very helpful.

6. Lecture Notes are printed by photo-offset from camera-ready typed copy provided by the editors. For this purpose Springer-Verlag provides editors with technical instructions for the preparation of manuscripts and these should be distributed to all contributing authors. Springer-Verlag can also, on request, supply stationery on which the prescribed typing area is outlined. Some homogeneity in the presentation of the contributions is desirable.

   Careful preparation of manuscripts will help keep production time short and ensure a satisfactory appearance of the finished book. The actual production of a Lecture Notes volume normally takes 6 -8 weeks.

   Manuscripts should be at least 100 pages long. The final version should include a table of contents.

7. Editors receive a total of 50 free copies of their volume for distribution to the contributing authors, but no royalties. (Unfortunately, no reprints of individual contributions can be supplied.) They are entitled to purchase further copies of their book for their personal use at a discount of 33 1/3%, other Springer mathematics books at a discount of 20% directly from Springer-Verlag.

   Commitment to publish is made by letter of intent rather than by signing a formal contract. Springer-Verlag secures the copyright for each volume.

Vol. 1117: D.J. Aldous, J.A. Ibragimov, J. Jacod, Ecole d'Été de Probabilités de Saint-Flour XIII - 1983. Édité par P.L. Hennequin. IX, 409 pages. 1985.

Vol. 1118: Grossissements de filtrations: exemples et applications. Seminaire, 1982/83. Edité par Th. Jeulin et M. Yor. V, 315 pages. 1985.

Vol. 1119: Recent Mathematical Methods in Dynamic Programming. Proceedings, 1984. Edited by I. Capuzzo Dolcetta, W.H. Fleming and T. Zolezzi. VI, 202 pages. 1985.

Vol. 1120: K. Jarosz, Perturbations of Banach Algebras. V, 118 pages. 1985.

Vol. 1121: Singularities and Constructive Methods for Their Treatment. Proceedings, 1983. Edited by P. Grisvard, W. Wendland and J.R. Whiteman. IX, 346 pages. 1985.

Vol. 1122: Number Theory. Proceedings, 1984. Edited by K. Alladi. VII, 217 pages. 1985.

Vol. 1123: Séminaire de Probabilités XIX 1983/84. Proceedings. Edité par J. Azéma et M. Yor. IV, 504 pages. 1985.

Vol. 1124: Algebraic Geometry, Sitges (Barcelona) 1983. Proceedings. Edited by E. Casas-Alvero, G.E. Welters and S. Xambó-Descamps. XI, 416 pages. 1985.

Vol. 1125: Dynamical Systems and Bifurcations. Proceedings, 1984. Edited by B.L.J. Braaksma, H.W. Broer and F. Takens. V, 129 pages. 1985.

Vol. 1126: Algebraic and Geometric Topology. Proceedings, 1983. Edited by A. Ranicki, N. Levitt and F. Quinn. V, 423 pages. 1985.

Vol. 1127: Numerical Methods in Fluid Dynamics. Seminar. Edited by F. Brezzi, VII, 333 pages. 1985.

Vol. 1128: J. Elschner, Singular Ordinary Differential Operators and Pseudodifferential Equations. 200 pages. 1985.

Vol. 1129: Numerical Analysis, Lancaster 1984. Proceedings. Edited by P.R. Turner. XIV, 179 pages. 1985.

Vol. 1130: Methods in Mathematical Logic. Proceedings, 1983. Edited by C.A. Di Prisco. VII, 407 pages. 1985.

Vol. 1131: K. Sundaresan, S. Swaminathan, Geometry and Nonlinear Analysis in Banach Spaces. III, 116 pages. 1985.

Vol. 1132: Operator Algebras and their Connections with Topology and Ergodic Theory. Proceedings, 1983. Edited by H. Araki, C.C. Moore, Ş. Strătilă and C. Voiculescu. VI, 594 pages. 1985.

Vol. 1133: K.C. Kiwiel, Methods of Descent for Nondifferentiable Optimization. VI, 362 pages. 1985.

Vol. 1134: G.P. Galdi, S. Rionero, Weighted Energy Methods in Fluid Dynamics and Elasticity. VII, 126 pages. 1985.

Vol. 1135: Number Theory, New York 1983–84. Seminar. Edited by D.V. Chudnovsky, G.V. Chudnovsky, H. Cohn and M.B. Nathanson. V, 283 pages. 1985.

Vol. 1136: Quantum Probability and Applications II. Proceedings, 1984. Edited by L. Accardi and W. von Waldenfels. VI, 534 pages. 1985.

Vol. 1137: Xiao G., Surfaces fibrées en courbes de genre deux. IX, 103 pages. 1985.

Vol. 1138: A. Ocneanu, Actions of Discrete Amenable Groups on von Neumann Algebras. V, 115 pages. 1985.

Vol. 1139: Differential Geometric Methods in Mathematical Physics. Proceedings, 1983. Edited by H. D. Doebner and J. D. Hennig. VI, 337 pages. 1985.

Vol. 1140: S. Donkin, Rational Representations of Algebraic Groups. VII, 254 pages. 1985.

Vol. 1141: Recursion Theory Week. Proceedings, 1984. Edited by H.-D. Ebbinghaus, G.H. Müller and G.E. Sacks. IX, 418 pages. 1985.

Vol. 1142: Orders and their Applications. Proceedings, 1984. Edited by I. Reiner and K. W. Roggenkamp. X, 306 pages. 1985.

Vol. 1143: A. Krieg, Modular Forms on Half-Spaces of Quaternions. XIII, 203 pages. 1985.

Vol. 1144: Knot Theory and Manifolds. Proceedings, 1983. Edited by D. Rolfsen. V, 163 pages. 1985.

Vol. 1145: G. Winkler, Choquet Order and Simplices. VI, 143 pages. 1985.

Vol. 1146: Séminaire d'Algèbre Paul Dubreil et Marie-Paule Malliavin. Proceedings, 1983–1984. Edité par M.-P. Malliavin. IV, 420 pages. 1985.

Vol. 1147: M. Wschebor, Surfaces Aléatoires. VII, 111 pages. 1985.

Vol. 1148: Mark A. Kon, Probability Distributions in Quantum Statistical Mechanics. V, 121 pages. 1985.

Vol. 1149: Universal Algebra and Lattice Theory. Proceedings, 1984. Edited by S. D. Comer. VI, 282 pages. 1985.

Vol. 1150: B. Kawohl, Rearrangements and Convexity of Level Sets in PDE. V, 136 pages. 1985.

Vol 1151: Ordinary and Partial Differential Equations. Proceedings, 1984. Edited by B.D. Sleeman and R.J. Jarvis. XIV, 357 pages. 1985.

Vol. 1152: H. Widom, Asymptotic Expansions for Pseudodifferential Operators on Bounded Domains. V, 150 pages. 1985.

Vol. 1153: Probability in Banach Spaces V. Proceedings, 1984. Edited by A. Beck, R. Dudley, M. Hahn, J. Kuelbs and M. Marcus. VI, 457 pages. 1985.

Vol. 1154: D.S. Naidu, A.K. Rao, Singular Pertubation Analysis of Discrete Control Systems. IX, 195 pages. 1985.

Vol. 1155: Stability Problems for Stochastic Models. Proceedings, 1984. Edited by V.V. Kalashnikov and V.M. Zolotarev. VI, 447 pages. 1985.

Vol. 1156: Global Differential Geometry and Global Analysis 1984. Proceedings, 1984. Edited by D. Ferus, R.B. Gardner, S. Helgason and U. Simon. V, 339 pages. 1985.

Vol. 1157: H. Levine, Classifying Immersions into $\mathbb{R}^4$ over Stable Maps of 3-Manifolds into $\mathbb{R}^2$. V, 163 pages. 1985.

Vol. 1158: Stochastic Processes – Mathematics and Physics. Proceedings, 1984. Edited by S. Albeverio, Ph. Blanchard and L. Streit. VI, 230 pages. 1986.

Vol. 1159: Schrödinger Operators, Como 1984. Seminar. Edited by S. Graffi. VIII, 272 pages. 1986.

Vol. 1160: J.-C. van der Meer, The Hamiltonian Hopf Bifurcation. VI, 115 pages. 1985.

Vol. 1161: Harmonic Mappings and Minimal Immersions, Montecatini 1984. Seminar. Edited by E. Giusti. VII, 285 pages. 1985.

Vol. 1162: S.J.L. van Eijndhoven, J. de Graaf, Trajectory Spaces, Generalized Functions and Unbounded Operators. IV, 272 pages. 1985.

Vol. 1163: Iteration Theory and its Functional Equations. Proceedings, 1984. Edited by R. Liedl, L. Reich and Gy. Targonski. VIII, 231 pages. 1985.

Vol. 1164: M. Meschiari, J.H. Rawnsley, S. Salamon, Geometry Seminar "Luigi Bianchi" II – 1984. Edited by E. Vesentini. VI, 224 pages. 1985.

Vol. 1165: Seminar on Deformations. Proceedings, 1982/84. Edited by J. Ławrynowicz. IX, 331 pages. 1985.

Vol. 1166: Banach Spaces. Proceedings, 1984. Edited by N. Kalton and E. Saab. VI, 199 pages. 1985.

Vol. 1167: Geometry and Topology. Proceedings, 1983–84. Edited by J. Alexander and J. Harer. VI, 292 pages. 1985.

Vol. 1168: S.S. Agaian, Hadamard Matrices and their Applications. III, 227 pages. 1985.

Vol. 1169: W.A. Light, E.W. Cheney, Approximation Theory in Tensor Product Spaces. VII, 157 pages. 1985.

Vol. 1170: B.S. Thomson, Real Functions. VII, 229 pages. 1985.

Vol. 1171: Polynômes Orthogonaux et Applications. Proceedings, 1984. Edité par C. Brezinski, A. Draux, A.P. Magnus, P. Maroni et A. Ronveaux. XXXVII, 584 pages. 1985.

Vol. 1172: Algebraic Topology, Göttingen 1984. Proceedings. Edited by L. Smith. VI, 209 pages. 1985.

Vol. 1173: H. Delfs, M. Knebusch, Locally Semialgebraic Spaces. XVI, 329 pages. 1985.

Vol. 1174: Categories in Continuum Physics, Buffalo 1982. Seminar. Edited by F.W. Lawvere and S.H. Schanuel. V, 126 pages. 1986.

Vol. 1175: K. Mathiak, Valuations of Skew Fields and Projective Hjelmslev Spaces. VII, 116 pages. 1986.

Vol. 1176: R.R. Bruner, J.P. May, J.E. McClure, M. Steinberger, H_∞ Ring Spectra and their Applications. VII, 388 pages. 1986.

Vol. 1177: Representation Theory I. Finite Dimensional Algebras. Proceedings, 1984. Edited by V. Dlab, P. Gabriel and G. Michler. XV, 340 pages. 1986.

Vol. 1178: Representation Theory II. Groups and Orders. Proceedings, 1984. Edited by V. Dlab, P. Gabriel and G. Michler. XV, 370 pages. 1986.

Vol. 1179: Shi J.-Y. The Kazhdan-Lusztig Cells in Certain Affine Weyl Groups. X, 307 pages. 1986.

Vol. 1180: R. Carmona, H. Kesten, J.B. Walsh, École d'Été de Probabilités de Saint-Flour XIV – 1984. Édité par P.L. Hennequin. X, 438 pages. 1986.

Vol. 1181: Buildings and the Geometry of Diagrams, Como 1984. Seminar. Edited by L. Rosati. VII, 277 pages. 1986.

Vol. 1182: S. Shelah, Around Classification Theory of Models. VII, 279 pages. 1986.

Vol. 1183: Algebra, Algebraic Topology and their Interactions. Proceedings, 1983. Edited by J.-E. Roos. XI, 396 pages. 1986.

Vol. 1184: W. Arendt, A. Grabosch, G. Greiner, U. Groh, H.P. Lotz, U. Moustakas, R. Nagel, F. Neubrander, U. Schlotterbeck, One-parameter Semigroups of Positive Operators. Edited by R. Nagel. X, 460 pages. 1986.

Vol. 1185: Group Theory, Beijing 1984. Proceedings. Edited by Tuan H.F. V, 403 pages. 1986.

Vol. 1186: Lyapunov Exponents. Proceedings, 1984. Edited by L. Arnold and V. Wihstutz. VI, 374 pages. 1986.

Vol. 1187: Y. Diers, Categories of Boolean Sheaves of Simple Algebras. VI, 168 pages. 1986.

Vol. 1188: Fonctions de Plusieurs Variables Complexes V. Séminaire, 1979–85. Edité par François Norguet. VI, 306 pages. 1986.

Vol. 1189: J. Lukeš, J. Malý, L. Zajíček, Fine Topology Methods in Real Analysis and Potential Theory. X, 472 pages. 1986.

Vol. 1190: Optimization and Related Fields. Proceedings, 1984. Edited by R. Conti, E. De Giorgi and F. Giannessi. VIII, 419 pages. 1986.

Vol. 1191: A.R. Its, V.Yu. Novokshenov, The Isomonodromic Deformation Method in the Theory of Painlevé Equations. IV, 313 pages. 1986.

Vol. 1192: Equadiff 6. Proceedings, 1985. Edited by J. Vosmansky and M. Zlámal. XXIII, 404 pages. 1986.

Vol. 1193: Geometrical and Statistical Aspects of Probability in Banach Spaces. Proceedings, 1985. Edited by X. Femique, B. Heinkel, M.B. Marcus and P.A. Meyer. IV, 128 pages. 1986.

Vol. 1194: Complex Analysis and Algebraic Geometry. Proceedings, 1985. Edited by H. Grauert. VI, 235 pages. 1986.

Vol. 1195: J.M. Barbosa, A.G. Colares, Minimal Surfaces in $\mathbb{R}^3$. X, 124 pages. 1986.

Vol. 1196: E. Casas-Alvero, S. Xambó-Descamps, The Enumerative Theory of Conics after Halphen. IX, 130 pages. 1986.

Vol. 1197: Ring Theory. Proceedings, 1985. Edited by F.M.J. van Oystaeyen. V, 231 pages. 1986.

Vol. 1198: Séminaire d'Analyse, P. Lelong – P. Dolbeault – H. Skoda. Seminar 1983/84. X, 260 pages. 1986.

Vol. 1199: Analytic Theory of Continued Fractions II. Proceedings, 1985. Edited by W.J. Thron. VI, 299 pages. 1986.

Vol. 1200: V.D. Milman, G. Schechtman, Asymptotic Theory of Finite Dimensional Normed Spaces. With an Appendix by M. Gromov. VIII, 156 pages. 1986.

Vol. 1201: Curvature and Topology of Riemannian Manifolds. Proceedings, 1985. Edited by K. Shiohama, T. Sakai and T. Sunada. VII, 336 pages. 1986.

Vol. 1202: A. Dür, Möbius Functions, Incidence Algebras and Power Series Representations. XI, 134 pages. 1986.

Vol. 1203: Stochastic Processes and Their Applications. Proceedings, 1985. Edited by K. Itô and T. Hida. VI, 222 pages. 1986.

Vol. 1204: Séminaire de Probabilités XX, 1984/85. Proceedings. Edité par J. Azéma et M. Yor. V, 639 pages. 1986.

Vol. 1205: B.Z. Moroz, Analytic Arithmetic in Algebraic Number Fields. VII, 177 pages. 1986.

Vol. 1206: Probability and Analysis, Varenna (Como) 1985. Seminar. Edited by G. Letta and M. Pratelli. VIII, 280 pages. 1986.

Vol. 1207: P.H. Bérard, Spectral Geometry: Direct and Inverse Problems. With an Appendix by G. Besson. XIII, 272 pages. 1986.

Vol. 1208: S. Kaijser, J.W. Pelletier, Interpolation Functors and Duality. IV, 167 pages. 1986.

Vol. 1209: Differential Geometry, Peñíscola 1985. Proceedings. Edited by A.M. Naveira, A. Ferrández and F. Mascaró. VIII, 306 pages. 1986.

Vol. 1210: Probability Measures on Groups VIII. Proceedings, 1985. Edited by H. Heyer. X, 386 pages. 1986.

Vol. 1211: M.B. Sevryuk, Reversible Systems. V, 319 pages. 1986.

Vol. 1212: Stochastic Spatial Processes. Proceedings, 1984. Edited by P. Tautu. VIII, 311 pages. 1986.

Vol. 1213: L.G. Lewis, Jr., J.P. May, M. Steinberger, Equivariant Stable Homotopy Theory. IX, 538 pages. 1986.

Vol. 1214: Global Analysis – Studies and Applications II. Edited by Yu.G. Borisovich and Yu.E. Gliklikh. V, 275 pages. 1986.

Vol. 1215: Lectures in Probability and Statistics. Edited by G. del Pino and R. Rebolledo. V, 491 pages. 1986.

Vol. 1216: J. Kogan, Bifurcation of Extremals in Optimal Control. VIII, 106 pages. 1986.

Vol. 1217: Transformation Groups. Proceedings, 1985. Edited by S. Jackowski and K. Pawalowski. X, 396 pages. 1986.

Vol. 1218: Schrödinger Operators, Aarhus 1985. Seminar. Edited by E. Balslev. V, 222 pages. 1986.

Vol. 1219: R. Weissauer, Stabile Modulformen und Eisensteinreihen. III, 147 Seiten. 1986.

Vol. 1220: Séminaire d'Algèbre Paul Dubreil et Marie-Paule Malliavin. Proceedings, 1985. Edité par M.-P. Malliavin. IV, 200 pages. 1986.

Vol. 1221: Probability and Banach Spaces. Proceedings, 1985. Edited by J. Bastero and M. San Miguel. XI, 222 pages. 1986.

Vol. 1222: A. Katok, J.-M. Strelcyn, with the collaboration of F. Ledrappier and F. Przytycki, Invariant Manifolds, Entropy and Billiards; Smooth Maps with Singularities. VIII, 283 pages. 1986.

Vol. 1223: Differential Equations in Banach Spaces. Proceedings, 1985. Edited by A. Favini and E. Obrecht. VIII, 299 pages. 1986.

Vol. 1224: Nonlinear Diffusion Problems, Montecatini Terme 1985. Seminar. Edited by A. Fasano and M. Primicerio. VIII, 188 pages. 1986.

Vol. 1225: Inverse Problems, Montecatini Terme 1986. Seminar. Edited by G. Talenti. VIII, 204 pages. 1986.

Vol. 1226: A. Buium, Differential Function Fields and Moduli of Algebraic Varieties. IX, 146 pages. 1986.

Vol. 1227: H. Helson, The Spectral Theorem. VI, 104 pages. 1986.

Vol. 1228: Multigrid Methods II. Proceedings, 1985. Edited by W. Hackbusch and U. Trottenberg. VI, 336 pages. 1986.

Vol. 1229: O. Bratteli, Derivations, Dissipations and Group Actions on C*-algebras. IV, 277 pages. 1986.

Vol. 1230: Numerical Analysis. Proceedings, 1984. Edited by J.-P. Hennart. X, 234 pages. 1986.

Vol. 1231: E.-U. Gekeler, Drinfeld Modular Curves. XIV, 107 pages. 1986.